LOGICISM AND ITS PHILOSOPHICAL LEGACY

The idea that mathematics is reducible to logic has a long history, but it was Frege who gave logicism an articulation and defense that transformed it into a distinctive philosophical thesis with a profound influence on the development of philosophy in the twentieth century. This volume of classic, revised, and newly written essays by William Demopoulos examines logicism's principal legacy for philosophy: its elaboration of notions of analysis and reconstruction. The essays reflect on the deployment of these ideas by the principal figures in the history of the subject – Frege, Russell, Ramsey, and Carnap – and in doing so illuminate current concerns about the nature of mathematical and theoretical knowledge. Issues addressed include the nature of arithmetical knowledge in the light of Frege's theorem; the status of realism about the theoretical entities of physics; and the proper interpretation of empirical theories that postulate abstract structural constraints.

WILLIAM DEMOPOULOS is Professor of Philosophy at the University of Western Ontario. He has edited many volumes on foundational and philosophical issues arising in the sciences, including *Physical Theory and its Interpretation* (with Itamar Pitowsky in 2006), and *Frege's Philosophy of Mathematics* (1995).

LOGICISM AND ITS PHILOSOPHICAL LEGACY

WILLIAM DEMOPOULOS

CAMBRIDGE
UNIVERSITY PRESS

CAMBRIDGE UNIVERSITY PRESS
Cambridge, New York, Melbourne, Madrid, Cape Town,
Singapore, São Paulo, Delhi, Mexico City

Cambridge University Press
The Edinburgh Building, Cambridge CB2 8RU, UK

Published in the United States of America by Cambridge University Press, New York

www.cambridge.org
Information on this title: www.cambridge.org/9781107029804

First published 2013

Printed and bound in the United Kingdom by the MPG Books Group

A catalogue record for this publication is available from the British Library

Library of Congress Cataloguing in Publication data
Demopoulos, William.
Logicism and its philosophical legacy / William Demopoulos.
p.   cm.
Includes bibliographical references (p. 255) and index.
ISBN 978-1-107-02980-4
1. Logic, Symbolic and mathematical.   2. Mathematics – Philosophy.   I. Title.
QA9.D38   2013
511.3–dc23
2012023164

ISBN 978-1-107-02980-4 Hardback

*Dedicated to Rai, Billy, and Alexis, for whom my gratitude
is beyond expression*

Contents

Preface   *page* ix

   Introduction   1

1 Frege's analysis of arithmetical knowledge   8

2 Carnap's thesis   28

3 On extending "Empiricism, semantics and ontology" to the
    realism–instrumentalism controversy   46

4 Carnap's analysis of realism   68

5 Bertrand Russell's *The Analysis of Matter*: its historical context
    and contemporary interest (with Michael Friedman)   90

6 On the rational reconstruction of our theoretical knowledge   108

7 Three views of theoretical knowledge   140

8 Frege and the rigorization of analysis   169

9 The philosophical basis of our knowledge of number   185

10 The 1910 *Principia*'s theory of functions and classes   210

11 Ramsey's extensional propositional functions   238

Bibliography   255
Index   266

vii

Preface

My appreciation of Frege, like my love of opera, is something I acquired relatively late in life. I did not read *Grundlagen* until long after I had left graduate school, and I came to Frege by a circuitous route. I was motivated by Hilary Putnam's citation of Michael Dummett's elegant characterization of realism to look at Dummett's great work, *Frege: Philosophy of Language*. This was a book I approached with cautious interest, reading it on visits to Heffers bookstore during the summer of 1975 – the summer of the beginning of the "great drought" and the summer of Arthur Ashe's win at Wimbledon. I was in Cambridge with my family where we lived in the flat in Mrs. Turner's House at Girton Corner. The house was situated on the edge of Trinity College Farm, and there was a glorious walk into town that took you through the farm, and, for part of the way, on an old Roman road. Another path led past the American Cemetery to Madingley and an iconic English pub with a thatched roof.

It was from Dummett's book that I gradually made my way to the serious study of Frege, and eventually to *Grundlagen*. The character of *Frege: Philosophy of Language* was unique, with its remarkable blend of historical and systematic themes. Dummett's practice of historically situating problems while maintaining a focus on topical issues made a great impression on me. So did his ability to read Frege creatively, and, as I later came to believe, without distortion. Dummett was important in another respect. Many years later I sent him my first paper on Frege (Chapter 8 of this collection) with a brief note, and no expectation of a reply. Dummett replied with a letter of several pages, summarizing and extending the argument of my paper. The letter was warm and generous, and it began a long correspondence and friendship that sustained my decision to work on Frege and the tradition he started.

Russell attracted me from the very beginning of my study of philosophy. I thought of Russell as a contemporary figure. He was still alive when I began to read him and he had published *My Philosophical Development* only

a few years before. My initial interest in Russell was focused less on his
philosophy of mathematics than his theory of knowledge, especially his
theory of our knowledge of the material world. In my final term as an
undergraduate at the University of Minnesota I attended Grover Maxwell's
graduate seminar on the meaning of theoretical terms. It was in this seminar
that Maxwell first formulated his conception of the significance of Russell's
later philosophy for what Grover called "the theory of theories." Being the
only undergraduate in the class was a heady experience. For my seminar
presentation I was asked to report on Herbert Feigl's prescient "Existential
hypotheses"; I learned a great deal in this seminar about the positivist and
post-positivist view of theories.

The University of Minnesota was the home of the Minnesota Center for
Philosophy of Science. About a year after becoming a graduate student at
Minnesota, I was able to participate in Center discussions and to make use
of some of the Center's archival resources. These included typed transcripts
of Center meetings that went back to the 1950s, some of them attended by
Carnap.

Grover came as close as anyone I have known to the Platonic ideal of a
teacher: knowledgeable without pretension; encouraging, but not uncrit-
ical; personally warm and utterly honest. His early death was a great loss,
and I have always regretted that my first serious study of Russell was
written only after Grover died. The paper was a contribution to a memorial conference for Grover that was organized by his Center colleagues. I
drove to the conference with my family along a route that took us first to
Chicago where we spent a short time with Michael Friedman. Michael,
already a close friend of many years, had been asked to comment on my
paper, and during this visit we discussed the paper and his comments.
Then we all continued on to Minneapolis together. I had never before
written a paper like the one for Grover's memorial: I hoped it would be of
interest to anyone concerned with the history of the subject, but I did not
think of it as history in the strict sense. At the same time, it attempted to
make a contribution to problems of current interest – problems about the
nature of knowledge, especially theoretical knowledge. On reflection, the
form of the paper was clearly influenced by my reading of Dummett, but I
did not recognize this at the time. Friedman's comments led naturally to
our joint paper which appears here as Chapter 5.

Apropos my remark about Grover and my failure to write anything on
Russell's views until after Grover's death, I should mention that I formed
the idea of writing a dissertation on Russell and the psychology of perception when I was a graduate student at Pittsburgh. I even went so far as to

write M. H. A. Newman in order to gain access to his correspondence with Russell when I learned that Russell's papers had recently been acquired by McMaster University. I had a brief exchange with Newman about Russell and his now well-known criticisms of Russell's views. But being young and foolish, I did not immediately appreciate the correctness of what he wrote me. Many years later Jack Copeland introduced me (by correspondence) to Newman's son, William Newman, who sent me his elegant and moving memoir about his father, together with a fine reproduction of Russell's gracious letter to Max Newman. The letter was for many years proudly displayed on the wall of the Newmans' living room.

Of the three figures with whose work these essays are principally concerned, my interest in Carnap was the slowest to develop. Even the enthusiasm of Michael Friedman was not enough to inspire an appropriately deep interest in Carnap. It was only the persistence of my former student, the late Graham Solomon, and our conversations about his series of papers on Carnap, jointly authored with Dave DeVidi, that forced me to take notice and develop the interest that I should have acquired much earlier. This decision has since been confirmed many times over. Serious students are drawn by the transparency of Carnap's intellectual integrity to uncover the large – and even sweeping – conception of philosophy that lies just below the surface of a deceptively dry presentation.

My work on the essays included here spans a period of thirty years. It has been facilitated by the intellectual stimulation of numerous friends, but there are some to whom I owe a special debt of gratitude. Michael Friedman is one of my oldest friends; Michael has been enormously generous with his knowledge of the tradition from Kant to Carnap, sound in his judgment, and unstinting in his encouragement. Anil Gupta has been an indispensable philosophical guide; I cannot overstate how much I have benefited from his critical acumen and profound grasp of current work in logic, epistemology, and philosophy of language. Robert DiSalle has been an exemplary colleague: unfailingly perceptive as an interlocutor; authoritative in his field; constant in his support. It was Robert who provided the impetus for this volume. My philosophical friendship of longest standing is my friendship with Jeffrey Bub, a friendship that began when we met at the Minnesota Center for Philosophy of Science and has continued with only minor interruptions ever since. My intellectual orientation was defined by our discussions about the philosophy and foundations of physics. The present volume benefited directly from Jeff when he and my highly accomplished former student, Melanie Frappier, provided me with a superb translation of the late essay of Poincaré that figures prominently in the discussion of Chapter 4.

Many people have influenced the final form of one or more of these essays. I have done my best to acknowledge influences at appropriate points in the text and in the acknowledgments sections which follow most of the essays, but I am sure to have failed to remember everyone. My apologies to anyone I have forgotten.

Whatever success attaches to the essays collected here would not have been possible without the benefit of working in a well-run department. This was my experience during those years when Robert Butts and Ausonio Marras chaired the Philosophy Department at Western. I consider myself very fortunate to have experienced a stewardship such as theirs.

I wish to thank the Social Sciences and Humanities Research Council of Canada and its innovative program of modest research time stipends for its support. Many of the essays in this volume were either conceived or written during my tenure as a Killam Fellow. It is a pleasure to acknowledge the support of the Killam Trusts; news that I was to be awarded one of its research fellowships came at a critical moment in my life.

In connection with the final preparation of this book, Ryan Samaroo worked very hard on the index and made numerous helpful suggestions. Sona Ghosh provided the fine formatting of the two diagrams. Last but far from least, I cannot imagine a better editor than Hilary Gaskin, my editor at Cambridge University Press.

Publication details for a previously published essay are given in the first footnote of the chapter in which the essay appears. My thanks to the publishers, journal editors, and volume editors for permission to reprint these essays here.

Introduction

The idea that mathematics is reducible to logic has a long history. But it was Frege who gave logicism an articulation and defense that transformed it into a distinctive philosophical thesis of major consequence. With Frege's intervention, the doctrine came to have a profound influence on the development of philosophy in the twentieth century: it led directly to modern second-order logic – essentially the product of Frege's analysis of the concept of generality – and to notions of analysis that came to define philosophical thinking in Britain, on the continent, and in America.

The recent revival of interest in Frege's philosophy of arithmetic has shown it to be of much more than just historical interest. This revival was occasioned by Crispin Wright's important discovery that a great deal of Frege's theory of number is unaffected by the contradiction in his theory of classes. The reevaluation of Frege inspired by Wright's discovery has spawned a large philosophical and technical literature. Though informed by the technical results recorded in this literature, the goal of the essays collected here is general and philosophical. In addition to their contribution to the reevaluation of Frege's philosophy of mathematics, they argue that Frege's work contains insights that bear on fundamental problems of interpretation that arise in the context of empirical theories. The essays offer a series of critical reflections on a variety of applications of different conceptions of analysis – all of them having their source in the logicist tradition – to the philosophy of mathematics and the philosophy of science. An assumption they attempt to encourage is that there is a great deal still to be learned about the nature of analysis and how it was practiced by Frege and the principal historical figures in the logicist tradition who followed him – Russell, Ramsey, and Carnap.

While I hope that the essays in this volume will be of interest to historians, especially historians of the twentieth-century analytic tradition, they are not – nor do they pretend to be – scholarly historical studies. I have tried to exhibit the main features of the variety of alternative analytical

I

approaches that different logicists have pursued in their separate attempts to define what is distinctive about arithmetical knowledge. And I have been especially concerned to isolate what was characteristic of Frege's contribution to the analysis of our knowledge of number, and to illuminate, by comparison with his contribution, the analyses of others. But I have not hesitated to emphasize a reconstruction of an author's view if I thought doing so was clarifying or suggestive of a more successful approach to the problem I was considering; I have proceeded in this way even when such a reconstruction represented an obvious departure from the view that inspired it. This is particularly true of my accounts of Frege's analysis of arithmetical knowledge in Chapter 1, Carnap's elaboration and defense of the thesis that mathematics is not factual in Chapter 2, and *Principia*'s theory of functions and classes in Chapter 10.

Frege's criterion of identity for number occupies a prominent position in these essays. The criterion was put forward in *Grundlagen* as an "informative" answer to the question, 'Under what conditions should we regard statements of the form, "The number of *F*s is the same as the number of *G*s," true?' According to Frege, such statements express "recognition judgments": they express our recognition that the same number has been presented in two different ways, as the number of one or another concept. The criterion of identity asserts that a recognition judgment is true if and only if its constituent concepts are in one–one correspondence.

In the secondary literature, Frege's criterion of identity is sometimes referred to as a partial contextual definition of the truth conditions for numerical statements of identity. Setting aside the notion of a contextual definition, the definition the criterion expresses is said to be *partial* because it covers only some identities involving numerical singular terms. It covers those cases where the singular terms are of the form 'the number of . . .,' or, as with numeral names, are definable in terms of expressions of this form, but it leaves out cases where at least one of the singular terms is not of this character. However, even if the criterion of identity is in this sense only a partial contextual definition, when taken as an axiom or principle, it suffices for a mathematically adequate theory of number – one that gives rise to an analysis of arithmetical knowledge of considerable philosophical interest.

In the essays that follow, Frege's criterion of identity is discussed in connection with two rather different applications of it. It is discussed first in Chapter 1 in connection with the theory of arithmetical knowledge, where I formulate a theory that is *close* to Frege's intended account and compare it with the theory of Whitehead and Russell. In Chapter 2 the criterion is discussed in connection with Carnap, where it is applied to the problem of distinguishing between formal and factual components of the language of science. The essays of both chapters are new to this volume, but since the second application of Frege's criterion of identity is likely to be a much less familiar kind of application of the criterion than the first, it might be worthwhile to comment further on it here.

Hume had the idea that the difference between the epistemic status of arithmetic and geometry can be traced to the different criteria of identity that govern their fundamental notions – the notions of number and length, respectively. Indeed, when in sect. 63 of *Grundlagen*, Frege introduces the criterion of identity, he quotes from Hume's *Treatise*:

We are possest of a precise standard, by which we can judge of the equality and proportion of two numbers; and according as they correspond or not to that standard, we determine their relations without any possibility of error. When two numbers are so combin'd as that one has always an unite answering to every unite of the other, we pronounce them equal[.] (Book I, Part III, sect. I, para. 5)

That Frege quotes Hume in the course of introducing his criterion of identity for number led George Boolos to coin the name 'Hume's principle' for the criterion of identity. Motivated in no small measure by the attractiveness of Boolos's essays on the topic, the name has become established in the secondary literature devoted to the reevaluation and revival of Frege's contributions to the philosophy of mathematics. Whether or not it was intended, Boolos's terminology suggests an important change in perspective: it focuses attention on the role of the criterion of identity as an independently motivated axiom, rather than a definition that is incomplete or partial.

It is interesting to note that Frege does not quote Hume in full, but leaves out the remark with which the passage concludes:

and 'tis for want of such a standard of equality in extension, that geometry can scarce be esteem'd a perfect and infallible science.

Frege had little interest in Hume's application of criteria of identity – "standards of equality" in Hume's phrase – to the problem of elucidating the epistemological differences between arithmetic and geometry. But he certainly hoped to gain acceptance for Hume's principle, given its pivotal position in the central mathematical argument of *Grundlagen*: the entire development of the theory of number is based on it, and Frege may well have supposed that by citing the precedent of Hume, he could enhance the plausibility of basing his theory on this criterion. Nevertheless, Frege's favorable citation of Hume obscures important differences in their understanding of the criterion of identity: it is clear from the passage Frege quotes that Hume conflates the notion of a class with that of a plurality, and that he treats a plurality of things as a "number," rather than treating numbers as separate from – and representative of – the concepts or classes to which they "belong." Each of these differences points to a characteristic that is a hallmark of Frege's philosophy of mathematics.

The context of Hume's introduction of the standard of equality for number is actually *geometric*, and arithmetic enters the discussion only coincidentally. Hume observed that the idea that space is infinitely divisible cannot be based on perception or imagination since reflection quickly reveals that any division we effect must terminate after some indefinite but finite number of steps. The terminus of any process of division is not susceptible to further division, and although the termini are in this sense unextended spatial minima, they are minima that are reached after finitely many divisions. Against the suggestion that two spatial intervals are equal in length if they are composed of the same number of spatial minima, Hume argues that our idea of an interval is always of one for which the minima are "confounded" with one another. This confounding has the consequence that it is impossible to base a notion of sameness of length on a one–one correspondence between the minima belonging to the two intervals. The situation is altogether different in the case of our judgments of sameness of number, for these concern pluralities of "unites" which we can clearly grasp as separate from one another. Hence the question whether the units of the pluralities to which arithmetic is applied are or are not in one–one correspondence has a clear sense and a determinate answer. But in the case of geometry, we must be content with the usual criterion for sameness of length, namely the coincidence of the end-points of the intervals being compared: in other words, the geometrical situation is just as it would be if we were dealing with a continuum, where comparisons are necessarily approximate. Geometry is therefore fallible and "imperfect" in a way that arithmetic is not.[1]

In Chapter 2 I argue that there is a sense in which the question, 'What distinguishes applied geometry from applied arithmetic?' is illuminated by an examination of the criteria of identity that are appropriate to arithmetical and geometric notions. Although Hume must be credited for first raising

---

[1] Notice that the appropriateness of the arithmetical criterion in the context of a discrete space can be reasonably questioned even if one does not accept the relevance of Hume's contention that spatial minima are necessarily "confounded" by us. It would not be unreasonable to reject the arithmetical criterion for sameness of length if the judgments of congruence to which it led conflicted with those that are forced by our customary geometric procedures for establishing congruence.

the possibility that differences associated with their criteria of identity bear on the difference in epistemic status we attribute to applied geometry and applied arithmetic, his positive proposals fail to isolate the correct point. Contrary to Hume, the difference is not that one standard is precise in a way that the other can never be, or that one enjoys a kind of infallibility that the other can never attain. Rather the central difference is that the criteria of identity for geometrical notions like *length* are empirically constrained in a way that the criterion of identity for *number* is not. The manner in which geometrical notions are differently constrained depends on methodological considerations of some subtlety, requiring for their full appreciation the discussion of chronometry and the criterion of identity for *time of occurrence*. The chapter concludes by showing how these observations can be exploited to support Carnap's claim that applied geometry has a factual content that applied arithmetic lacks. This claim lies at the center of Carnap's insistence that an adequate reconstruction of physics must include a principled distinction between its factual and nonfactual components.

There is an interesting conceptual difference between Frege and Hume which would have been clear to Frege, but would have fallen outside Hume's conceptual horizon. This difference is attributable to their divergent views on the scope of logic. Although a small point, it leads naturally to an observation about Russell and a major preoccupation of several of the essays collected here. Hume's discussion of infinite divisibility raises two questions: (i) 'What is the basis for our belief that space is infinitely divisible?' and (ii) 'How do we come to the concept of infinite divisibility?' Hume treats the first question as one about space as we perceive or imagine it; he then proceeds to address the question on the basis of the phenomenological evidence that is provided by our perception and imagination. As for the second question, Frege had the conceptual resources from which he could argue that whether or not the basis for our *belief* in infinite divisibility is decided by what is allowed by our perceptual and imaginative capacities, our *concept* of infinite divisibility does not rest on perception or imagination *because infinite divisibility – i.e. denseness – is a purely logical notion.* Even if Hume were to agree with the conclusion that our concept of infinite divisibility does not rest on our perception or imagination, he could not have advanced this argument on its behalf.

Similarly, independently of whether the basic laws of arithmetic are reducible to the laws of logic, it was a discovery of some interest that the truth of a recognition judgment should depend on a condition that is characterizable in terms drawn exclusively from logic. That this notion of similarity of properties or Fregean concepts is a wholly logical one, and therefore a notion of similarity that does not depend on experience or "a priori intuition," was the insight that led Russell to the view that the notion of *structural similarity* – which is a simple extension of one–one correspondence – might be capable of resolving classical metaphysical and epistemological problems about the relation of appearance to reality. The origins of Russell's "structuralism" therefore trace back to ideas that were integral to the development of logicism. Chapter 4 contains an extended introduction to the essays that deal with Russell's structuralism and its elaboration in the work of Ramsey, Carnap, and others, and its initial sections should be consulted for an overview of the topics these essays address.

Although the order of the chapters is the order in which I believe the essays are best read, each essay stands by itself and can be read independently of the others. I should, however, note that the previously unpublished essays were written not only to advance new ideas, but to introduce and orient the reader to the older essays in this collection. Indeed, the first four essays have a particularly strong claim to being read before the others and in the order in which they are presented, and they come close to forming a monograph of their own. As I mentioned earlier, Chapter 1 refines my view of what is of contemporary interest in Frege's theory of number and presents the background to Chapter 2; Chapter 1 also serves as an introduction to all the essays that are specifically concerned with logicism and the philosophy of mathematics, namely Chapters 8–11. Chapter 2 is the chapter most explicitly concerned with applying the central ideas of Frege's logicism to issues in the philosophy of science, and it forms a natural transition from Chapter 1 to Chapters 3 and 4. Chapter 3 addresses whether someone sympathetic to Carnap's views on questions of realism in the philosophy of mathematics is committed to a similar view of such questions in the philosophy of physics. In particular, it considers whether his distinction between internal and external questions can be applied differentially to questions about the existence of abstract entities and questions about the existence of the theoretical entities of physics. Chapter 4 is not only an introduction to Chapters 5–7; it also connects the discussion of Carnap's distinction between internal and external questions with issues raised by "structuralist" theories of theoretical knowledge, and it complements and extends the earlier discussion in Chapter 3 of the reality of theoretical entities.

In general, I have made only stylistic changes or changes for the purpose of clarification to the previously published essays. The major exception is Chapter 10 which, in addition to many such changes, has been significantly expanded by the addition of an appendix on Russell's propositional paradox, its extension to Fregean thoughts, and its relevance to Dedekind's theory of our knowledge of number. Where I have found a formulation or point of view with which I no longer agree, I have, in cases where it seemed important to do so, indicated this by a new footnote; but I have refrained from rewriting passages to reflect my current view. All footnotes that record such disagreements or direct the reader to a subsequent discussion use a special mark and are annotated as having been added in 2012.

CHAPTER I

Frege's analysis of arithmetical knowledge

Philosophy confines itself to universal concepts; mathematics can achieve nothing by concepts alone but hastens at once to intuition, in which it considers the concept *in concreto*, though not empirically, but only in an intuition which it presents *a priori*, that is, which it has constructed, and in which whatever follows from the universal conditions of the construction must be universally valid of the object thus constructed.

(Kant 1787, A716/B744)

Frege most directly engages Kant when, in *Grundlagen*,[1] he presents his formulations of the analytic–synthetic and a priori–a posteriori distinctions. The discussion of Frege's account of these distinctions has usually focused on whether the conception of logic which informs his definition of analyticity undermines his claim to have shown that arithmetic is analytic in a sense that Kant would have been concerned to deny. I will argue that Frege has an elegant explanation of the *apriority* of arithmetic, one that challenges Kant even if Frege's claim to have reduced arithmetic to logic is rejected. In the course of discussing Frege's explanation of the apriority of arithmetic, I will also clarify certain fundamental differences between his and Whitehead and Russell's theories of number – differences which bear importantly on their respective accounts of the nature of arithmetical knowledge. I will show that even if Frege's understanding of Kant was defective in every detail, knowing what he took himself to be reacting against and correcting in Kant's philosophy of arithmetic is of interest for the light it casts on the development of logicism.

---

[1] Frege (1884). My citations of Frege (1879), (1884), and (1893/1903) – his *Begriffsschrift*, *Grundlagen*, and *Grundgesetze*, respectively – use the mnemonic abbreviations, *Bg*, *Gl*, and *Gg*, followed by the relevant page or section number.

I.I FREGE'S INTEREST IN RIGOR

It may seem obvious that Frege's interest in rigor – his interest in providing a framework in which it would be possible to cast proofs in a canonical gap-free form – is driven by the problem of providing a proper justification for believing the truth of the propositions of arithmetic. One can easily find quotations from Frege which show that he sometimes at least wrote as if he took his task to be one of securing arithmetic; and it would be foolish to deny that the goals of cogency and consistency were an important part of the nineteenth-century mathematical tradition of which Frege was a part. Nevertheless, I think it can be questioned how far worries about the consistency or cogency of mathematics, generated perhaps by a certain incompleteness of its arguments, were motivating factors for Frege's logicism or for the other foundational investigations of the period.

There is another, largely neglected, component to Frege's concern with rigor that not only has an intrinsic interest, but also elucidates his views on the nature and significance of intuition in mathematical proof and his conception of his mathematical and foundational accomplishments. According to this component, Frege's concern with rigor is predominantly motivated by his desire to show that arithmetic does not depend on Kantian intuition, a concern Frege inherited from the tradition in analysis initiated by Cauchy and Bolzano, and carried forward by Weierstrass, Cantor, and Dedekind.

Shortly after the publication of *Begriffsschrift* Frege wrote a long study of its relationship to Boole's logical calculus.[2] The paper carries out a detailed proof (in the notation of *Begriffsschrift*) of the theorem that the sum of two multiples of a number is a multiple of that number. In addition to the laws and rules of inference of *Begriffsschrift*, Frege appeals only to the associativity of addition and to the fact that zero is a right identity with respect to addition. He avoids the use of mathematical induction by applying his definition of *following in a sequence* to the case of the number series. The paper also includes definitions of a number of elementary concepts of analysis (again in the notation of *Begriffsschrift*). It has been insufficiently emphasized that neither in this paper nor in *Begriffsschrift* does Frege suggest that the arithmetical theorems proved there are not correctly regarded as self-evident, or that without a *Begriffsschrift*-style proof, they and the propositions that depend on them might reasonably be doubted. Frege's point is rather that without gap-free proofs one might be misled into

---

[2] Frege (1880/1), published only posthumously.

thinking that arithmetical reasoning is based on intuition. As he puts the matter in the introductory paragraph to Part III of *Begriffsschrift*:

Throughout the present [study] we see how pure thought, irrespective of any content given by the senses or even by an intuition *a priori*, can, solely from the content that results from its own constitution, bring forth judgements that at first sight appear to be possible only on the basis of some intuition.

The same point is made in *Grundlagen* when, near the end of the work (sects. 90–1), Frege comments on *Begriffsschrift*. Frege is quite clear that the difficulty with gaps or jumps in the usual proofs of arithmetical propositions is not that they might hide an unwarranted or possibly false inference, but that their presence obscures the true character of the reasoning:

In proofs as we know them, progress is by jumps, which is why the variety of types of inference in mathematics appears to be so excessively rich; the correctness of such a transition is immediately self-evident to us; whereupon, since it does not obviously conform to any of the recognized types of logical inference, we are prepared to accept its self-evidence forthwith as intuitive, and the conclusion itself as a synthetic truth – and this even when obviously it holds good of much more than merely what can be intuited.

On these lines what is synthetic and based on intuition cannot be sharply separated from what is analytic . . .

To minimize these drawbacks, I invented my concept writing. It is designed to produce expressions which are shorter and easier to take in . . . so that no step is permitted which does not conform to the rules which are laid down once and for all. It is impossible, therefore, for any premiss to creep into a proof without being noticed. In this way I have, without borrowing any axiom from intuition, given a proof of a proposition[3] which might at first sight be taken for synthetic.

It might seem that to engage such passages it is necessary to enter into a detailed investigation of the Kantian concept of an a priori intuition. But to understand Frege's thought it is sufficient to recall that for the Kantian mathematical tradition of the period our a priori intuitions are of space and time, and the study of space and time falls within the provinces of geometry and kinematics. It follows that the dependence of a basic principle of arithmetic on some a priori intuition would imply that arithmetic lacks the autonomy and generality we associate with it. To establish its basic principles, we would have to appeal to our knowledge of space and time; and then arithmetical principles, like those expressing mathematical induction and various structural properties of the ancestral, would ultimately

---

[3] The proposition to which Frege refers is the last proposition proved in *Begriffsschrift* – Proposition 133 – which states that the ancestral of a many–one relation satisfies a restricted form of connectedness.

come to depend for their full justification on geometry and kinematics. Frege's point in the passages quoted is that even if it were possible to justify arithmetical principles in this way, it would be a mistake to suppose that such an external justification is either necessary or appropriate when a justification that is internal to arithmetic is available. From this perspective, the search for proofs, characteristic of a mathematical investigation into foundations of the sort Frege was engaged in, is not motivated by any uncertainty concerning basic principles and their justification, but by the absence of an argument which establishes their autonomy from geometrical and kinematical ideas. And autonomy is important since it is closely linked to the question of the generality of the principles of arithmetic.

On this reading of the allusion to Kantian intuition, the rigor of *Begriffsschrift* is required in order to show that arithmetic has no need of spatial or temporal notions. Indeed, Frege's concern with autonomy and independence suffices to explain the whole of his interest in combating the incursion of Kantian intuition into arithmetic. In this respect, Frege's intellectual motivation echoes that of the nineteenth-century analysts. Thus as early as 1817 Bolzano wrote: "the concepts of *time* and *motion* are just as foreign to general mathematics as the concept of *space*."[4] And over fifty years later Dedekind was equally emphatic:

For our immediate purpose, however, another property of the system 𝕽 [of real numbers] is still more important; it may be expressed by saying that the system 𝕽 forms a well-arranged domain of one dimension extending to infinity on two opposite sides. What is meant by this is sufficiently indicated by my use of expressions borrowed from geometric ideas; but just for this reason it will be necessary to bring out clearly the corresponding purely arithmetic properties in order to avoid even the appearance [that] arithmetic [is] in need of ideas foreign to it.[5]

It must be conceded that the issues raised by the foundations of analysis are more varied than those that arise in the case of the arithmetic of the natural numbers; in the case of the real numbers, the usual explanation of the purpose of rigor – that it is a guarantee of cogency and a hedge against inconsistency and incoherence – while not a complete account of the matter, is certainly a significant part of the story. But when we turn to Frege's primary concern about the domain of intuition – namely, the arithmetic of natural numbers – skepticism simply plays no role in any of his arguments against its use: Frege nowhere rejects intuition for fear that it is a potentially faulty guide to truth. Those few passages which suggest

---

[4] Russ (1980, p. 161). [5] Dedekind (1872, p. 5).

otherwise are invariably concerned with arithmetic in the broad sense, which includes real analysis. When we factor arithmetic by real analysis we also factor out doubts about cogency and consistency. What is left is a concern with autonomy and independence of the sort expressed by Bolzano and Dedekind in the passages just quoted.

We may briefly summarize our discussion of the historical situation as follows. The interest in rigor has both philosophical and mathematical aspects. To begin with, we require proofs internal to arithmetic. This idea is invariably presented as a prohibition against the incursion of spatial and temporal notions into demonstrations of the propositions of arithmetic. Finding proofs which are free of such notions is the mathematical aspect of rigorization. But the motivation underlying the foundational interest in rigor also has an important philosophical component which is broadly architectonic: the philosophical dimension to Frege's foundational program is to establish the independence of our knowledge of arithmetical principles from our knowledge of spatial and temporal notions.

I.2 THE SIGNIFICANCE OF A DERIVATION OF ARITHMETIC FROM LOGIC

Anyone familiar with the thesis that arithmetic is part of logic but with only a passing acquaintance with Frege's writings might expect that Frege was concerned to show that the necessity of arithmetic is inherited from the necessity of the laws of logic. But there are two considerations that argue against this expectation. First, Frege barely addresses the question of what characterizes a truth as logical, and when he does, necessity plays no role in his answer. Frege's great contribution to logic was his formulation of higher-order polyadic logic with mixed generality; but he contributed very little to our understanding of what constitutes a logical notion or a logical proposition and still less to our understanding of logical necessity. For Frege the laws of logic are distinguished by their universal applicability. Secondly, in so far as the laws of logic are for Frege necessary, they are "necessities of thought," a feature that is exhibited by the fact that they are presupposed by all the special sciences. But for Frege, this is a feature that is shared by the laws of arithmetic – whether or not they are derivable from logic. Because of logic's universality, a reduction of arithmetic to it would preserve the generality of arithmetic, but given Frege's conception of logic, it would shed no light on either arithmetic's necessity or its generality, since it is clear that these are properties arithmetic already enjoys.

The necessity of arithmetic is implicit in Frege's acknowledgment that it is a species of a priori knowledge. On Frege's definition of 'a priori' this means that arithmetic is susceptible of a justification solely on the basis of general laws that neither need nor admit of proof.[6] To say that a law neither needs nor admits of proof is to say that the law is warranted, and no premise of any purported proof of it is more warranted than the law itself. To the extent to which this implicitly assumes a notion of necessity, it is a wholly epistemic one. But while Frege's understanding of necessity is entirely epistemic, the contemporary concern with the necessity of arithmetic is characteristically directed toward the metaphysical necessity of arithmetical truths; this concern is altogether absent in Frege, and became prominent in the logicist tradition only much later, with Ramsey's celebrated 1925 essay on the foundations of mathematics.[7]

Frege's understanding of the epistemological significance of a derivation of arithmetic from logic is subtle. Certainly such a derivation would make it clear that arithmetic is not *synthetic* a priori, which is something Frege certainly sought to establish. But the derivation of arithmetic from logic is not needed for the simpler thesis that arithmetical knowledge is encompassed by Frege's definition of a priori knowledge. To suppose otherwise would be to imply that an appeal to logical principles is required because there are no self-warranting *arithmetical* principles to sustain arithmetic's apriority. This could only be maintained if arithmetical principles were less warranted than the basic laws of logic, or if they lacked the requisite generality. Discounting, for the moment, the second alternative, but supposing arithmetical principles to be less warranted than those of logic, the point of a derivation of arithmetic from logic would be to provide a justification for arithmetic. But Frege did not maintain that the basic laws of arithmetic – by which I mean the second-order Peano axioms (PA^2) – are significantly less warranted than those of his logic.[8]

It is important always to bear in mind that *Grundlagen* was not written in response to a "crisis" in the foundations of mathematics. *Grundlagen* seeks above all to illuminate the character of our knowledge of arithmetic and to

---

[6] Sect. 3 of *Grundlagen* appears to offer this only as a sufficient condition, not as a necessary and sufficient condition as a definition would require. Here I follow Burge (2005, pp. 359–60) who argues, convincingly in my view, that the condition is intended to be both necessary and sufficient.

[7] See Ramsey (1925b, sect. 1). Ramsey's essay is discussed at length in Chapter 11, below.

[8] I do not regard the equation of the basic laws of arithmetic with the second-order Peano axioms as at all tendentious. Each of the familiar Peano axioms or a very close analogue of it occurs in the course of the mathematical discussion of sects. 74–83 of *Grundlagen*, where the implicit logical context is the second-order logic of *Begriffsschrift* which, as customarily interpreted, assumes full comprehension. For additional considerations in favor of this equation, see Dummett (1991a, pp. 12–13).

address various misconceptions, most notably the Kantian misconception that arithmetic rests on intuitions given a priori. Frege says on more than one occasion that the primary goal of his logicism is not to secure arithmetic, but to expose the proper dependence relations of its truths on others. The early sections of *Grundlagen* are quite explicit in framing this general epistemological project, as the following passages illustrate:

The aim of proof is, in fact, not merely to place the truth of a proposition beyond all doubt, but also to afford us insight into the dependence of truths on one another. (*Gl*, sect. 2)

[T]he fundamental propositions of arithmetic should be proved ... with the utmost rigour; for only if every gap in the chain of deductions is eliminated with the greatest care can we say with certainty upon what primitive truths the proof depends. (*Gl*, sect. 4)

[I]t is above all Number which has to be either defined or recognized as indefinable. This is the point which the present work is meant to settle. On the outcome of this task will depend the decision as to the nature of the laws of arithmetic. (*Gl*, sect. 4)

Although a demonstration of the analyticity of arithmetic would indeed show that the basic laws of arithmetic can be justified by those of logic, the principal interest of such a derivation is what it would reveal regarding the dependence of arithmetical principles on logical laws. This would be a result of broadly epistemological interest, but its importance would not necessarily be that of providing a warrant where one is otherwise lacking or insufficient.

As for the idea that numbers are abstract objects, this also plays a relatively minor role in *Grundlagen*, and it is certainly not part of a desideratum by which to gauge a theory of arithmetical knowledge. Frege's emphasis is rarely on the positive claim that numbers are abstract objects, but is almost always a negative one to the effect that numbers are not ideas, not collections of units, not physical aggregates, not symbols, and most importantly, are neither founded on Kantian intuition nor the objects of Kantian intuition.

Aside from the thesis that numbers are objects – arguments to concepts of first level – the only positive claim of Frege's regarding the nature of numbers is that they are extensions of concepts, a claim which in *Grundlagen* has the character of a convenience (*Gl*, sects. 69 and 107); nor does he pause to explain extensions of concepts, choosing instead to assume that the notion is generally understood (footnote 1 to sect. 68). Frege's mature, post-*Grundlagen*, view of the characterization of numbers as classes or extensions of concepts is decidedly less casual. But that it too is almost exclusively focused on the epistemological role of classes – they facilitate the thesis that arithmetic is recoverable by analysis from our knowledge of logic – is clearly expressed in his correspondence with Russell:

I myself was long reluctant to recognize ... classes, but saw no other possibility of placing arithmetic on a logical foundation. But the question is, How do we apprehend logical objects? And I found no other answer to it than this, We apprehend them as extensions of concepts ... I have always been aware that there are difficulties connected with [classes]: but what other way is there?[9]

We should distinguish two roles that sound or truth-preserving derivations are capable of playing in a foundational investigation. Let us call derivations that play the first of these roles *proofs* (or *demonstrations*), and let us distinguish them from derivations that support *analyses*. *Proofs* are derivations which enhance the justification of what they establish by deriving them from more securely established truths in accordance with logically sound principles. *Analyses* are derivations which are advanced in order to clarify the logical dependency relations among propositions and concepts. The derivations involved in analyses do not enhance the warrant of the conclusion drawn, but display its basis in other truths. Given this distinction, the derivation of arithmetic from logic would not be advanced as a *proof* of arithmetic's basic laws if these laws were regarded as established, but in support of an *analysis* of them and their constituent concepts – most importantly, the concept of number. The general laws, on the basis of which a proposition is shown to be a priori, neither need nor admit of proof in the sense of 'proof' just explained. And although general laws may stand in need of an analysis, the derivation their analysis rests upon may well be one that does not add to their epistemic warrant. Unless this distinction or some equivalent of it is acknowledged, it is difficult to maintain that the basic laws of arithmetic neither need nor admit of proof while advancing the thesis that they are derivable from the basic laws of logic. There may be difficulties associated with the notion of neither needing nor admitting of proof, but they are not those of excluding the very possibility of logicism by conceding that arithmetic has basic laws that neither need nor admit of proof.

I.3 THE PROBLEM OF APRIORITY

Suppose we put to one side the matter of arithmetic's being analytic or synthetic. Does there remain a serious question concerning the mere apriority of its basic laws? This is actually a somewhat more delicate

---

[9] Frege to Russell, July 28, 1902, in McGuinness (1980, pp. 140–1).

question than our discussion so far would suggest. Recall that truths are, for Frege, a priori if they possess a justification exclusively on the basis of *general* laws which themselves neither need nor admit of proof. An unusual feature of this definition is that it makes no reference to experience. A possible explanation for this aspect of Frege's formulation is his adherence to the first of the three "fundamental principles" announced in the introduction to *Grundlagen*: always to separate sharply the psychological from the logical, the subjective from the objective. Standard explanations of apriority in terms of independence from experience have the potential for introducing just such a confusion – which is why apriority and innateness became so entangled in the traditional debate between rationalists and empiricists. Frege seems to have regarded Mill's views as the result of precisely the confusion that a definition in terms of general laws – rather than facts of experience – is intended to avoid. He therefore put forward a formulation which avoids even the appearance of raising a psychological question.

Although Frege's definition is in this respect nonstandard, it is easy to see that it subsumes the standard definition which requires of an a priori truth that it have a justification that is independent of experience: The justification of a fact of experience must ultimately rest on instances, either of the fact appealed to or of another adduced in support of it. But an appeal to instances requires mention of particular objects. Hence if a justification appeals to a fact of experience, it must appeal to a statement that concerns a specific object and that is therefore not a general law. Therefore, if a truth fails the usual test of apriority, it will fail Frege's test as well. However the converse is not true: a truth may fail Frege's test because its justification involves mention of a particular object, but there is nothing in Frege's definition to require that this object must be an object of experience. This inequivalence of Frege's definition with the standard one would point to a defect if it somehow precluded a positive answer to the question of the apriority of arithmetic. Does it?

Thus far I have only argued that for Frege the basic laws of arithmetic are not significantly less warranted than those of his logic. But are they completely general? In a provocative and historically rich paper, Tyler Burge (2005, sect. v) argues that Frege's account of apriority prevents him from counting all the Peano axioms as a priori. Let us set mathematical induction aside for the moment. Then the remaining axioms characterize the concept of a natural number as Dedekind-infinite, i.e. as in one-to-one correspondence with one of its proper subconcepts. And among these axioms, there is one in particular which, as Burge observes, fails the test of generality because it expresses a thought involving a particular object:

It is, of course, central to Frege's logicist project that truths about the numbers – which Frege certainly regarded as particular, determinate, formal objects (e.g. *Gl*, sects. 13 and 18) – are derivative from general logical truths ... [But s]uppose Frege is mistaken, and arithmetic is not derivable in an epistemically fruitful way from purely general truths. Suppose that arithmetic has the form it appears to have – a form that includes primitive singular intentional contents or propositions. For example, in the Peano axiomatization, arithmetic seems primitively to involve the thought that o is a number ... If some such knowledge is primitive – underived from general principles – then it counts as a posteriori on Frege's characterization. This would surely be a defect of the characterization. (Burge 2005, pp. 384–5)

Burge argues that it matters little that for Frege zero is to be recovered as a "logical object," and that for this reason it is arguably not in the same category as the particular objects the definition of a priori knowledge is intended to exclude. The problem is that such a defense of Frege makes his argument for the apriority of arithmetic depend on his inconsistent theory of extensions. Hence the notion of a logical object can take us no closer to an account of this fundamental fact – the fact of apriority – concerning our knowledge of arithmetic. Thus, Burge concludes, unless logicism can be sustained, Frege is without an account of the apriority of arithmetic.

There is, however, another way of seeing how the different components of *Grundlagen* fit together, one that yields a complete solution to what I will call *the problem of apriority*:

To explain the apriority of arithmetic in Frege's terms, i.e. in terms of an epistemically fruitful derivation from general laws which do not depend on the doctrine of logical objects or the truth of logicism.

The reconstruction of *Grundlagen* that suggests itself as a solution to this problem is so natural that it is surprising that it has not been proposed before. It is, however, a reconstruction, not an interpretation of Frege's views. Frege may have conceived of *Grundlagen* in the way I am about to explain, but there are at least two considerations that argue against such a supposition. First, Frege's main purpose in *Grundlagen* is to establish the analyticity of arithmetic; but on the proposed reconstruction, the goal of establishing arithmetic's apriority receives the same emphasis as the demonstration of its analyticity. Secondly, Frege frequently appeals to primitive truths and their natural order; by contrast the reconstruction I will propose uses only the notion of a basic law of logic or of arithmetic, and it uses both notions in a philosophically neutral sense. In particular the reconstruction does not assume that basic laws reflect a "natural order of primitive truths"; however, it does follow *Grundlagen* in *not* explaining *neither needing nor admitting of proof* in terms of self-evidence. There has been some confusion about the role of self-evidence in *Grundlagen* which it is important to clear-up.[10]

In *Grundlagen* (sect. 5) Frege considers the Kantian thesis that facts about particular numbers are grounded in intuition. His criticism of this thesis is subtle and hinges on a dialectical use of self-evidence in the premise that when a truth fails to qualify as self-evident, our knowledge of it cannot be intuitive. Frege argues that since facts about very large numbers are not self-evident, our knowledge of such facts cannot be intuitive and, hence, Kantian intuition cannot account for our knowledge of *every* fact about particular numbers. Here it is important to recognize that Frege is not proposing that the justification of primitive truths *must* rest on their self-evidence. Such a "justification" would be incompatible with Frege's rejection of psychologism. Rather than asserting that primitive truths are justified by their self-evidence, Frege is pointing to an internal tension in what he understands to be the Kantian view of the role of self-evidence in justification and its relation to intuitive knowledge.

Frege goes on to argue – independently of the considerations we have just reviewed – that even if facts about particular numbers were intuitive, and therefore self-evident, they would be unsuitable as a basis for our knowledge of arithmetic. The difficulty is that facts about particular numbers are infinitely numerous. Hence to take them as *the* primitive truths or first principles would "conflict with one of the requirements of reason, which must be able to embrace all first principles in a survey" (*Gl*, sect. 5). So whether or not facts about particular numbers are intuitively given and self-evident, they cannot exhaust the principles on which our arithmetical knowledge is based, and Frege concludes that the Kantian notion of intuition is at the very least not a complete guide to arithmetic's primitive truths.

The positive argument of *Grundlagen* divides into two parts. The first, and by far the more intricate argument, addresses the problem of apriority. The second argument, which I will ignore except in so far as it illuminates the argument for apriority, is directed at showing the basic laws of arithmetic to be analytic.

A principal premise of the argument for apriority – a premise which is established in *Grundlagen* – is Frege's theorem, i.e. the theorem that PA^2 is recoverable as a definitional extension of the second-order theory – called

---

[10] In this connection, see, for example, Jeshion (2001).

FA, for *Frege arithmetic* – whose sole nonlogical axiom is a formalization of the statement known in the recent secondary literature as *Hume's principle*:

For any concepts *F* and *G*, the number of *F*s is the same as the number of *G*s if, and only if, there is a one-to-one correspondence between the *F*s and the *G*s.

Frege introduces Hume's principle in the context of a discussion (*Gl*, sects. 62–3) of the necessity of providing a *criterion of identity* for number – an informative statement of the condition under which we should judge that the same number has been presented to us in two different ways, as the number of two different concepts. Hume's principle is advanced as such a criterion of identity – a criterion by which to assess such *recognition judgments* as:

The number of *F*s is the same as the number of *G*s.

It is essential to the understanding of Frege's proposed criterion of identity that the cardinality operator – 'the number of (. . .)' – be interpreted by a mapping from concepts to objects. But Hume's principle may otherwise be understood in a variety of ways: as a partial contextual definition of the concept of number, or of the cardinality operator; or we could follow a suggestion of Ricketts (1997, p. 92) and regard Hume's principle not as a definition of number – not even a contextual one – but a definition of the second-level relation of *equinumerosity* which holds of first-level concepts.[11] However we regard it, the criterion of identity possesses the generality that is required of the premise of an argument that seeks to establish the apriority of a known truth. But basing arithmetic on Hume's principle achieves more than its mere derivation from a principle that does not mention a particular object: it effects an account of the basic laws of *pure* arithmetic that reveals their basis in the principle which controls the *applications* we make of the numbers in our cardinality judgments. Since Hume's principle is an arithmetical rather than a logical principle, the derivation of PA^2 from FA is not the reductive analysis that logicism promised. Nonetheless, it is of considerable epistemological interest since, in addition to recovering PA^2 from a general law, the derivation of PA^2 from FA is based on an account of number which explains the peculiar generality that attaches to arithmetic: *in so far as the cardinality operator acts on concepts, arithmetic is represented as being as general in its scope and application as conceptual thought itself.*

The criterion of identity is the cornerstone of Frege's entire philosophy of arithmetic. The elegance of his account of the theory of the natural numbers is founded on his derivation of the Dedekind infinity of the natural

---

[11] I am assuming that all these construals acknowledge the existence and uniqueness assumptions which are implicit in the use of the cardinality operator as well as the operator's intended interpretation as a mapping from concepts to objects.

numbers from the second-order theory whose sole axiom is the criterion of identity. In *Principia Mathematica*, Whitehead and Russell (1910 and 1912) proceeded very differently. Contrary to conventional opinion, the difficulty with Whitehead and Russell's postulation of an "Axiom of Infinity" is not that it simply assumes what they set out to prove. For this it certainly does not do. *Principia*'s Axiom of Infinity asserts the existence of what Russell called a *noninductive* class of individuals: a class, the number of whose elements is not given by any *inductive* number, where the inductive numbers consist of zero and its progeny with respect to the relation of immediate successor. A *reflexive class* is Russell's term for a Dedekind-infinite class. Russell was well aware of the fact that the proof that every noninductive class is reflexive depends on the Axiom of Choice. Russell did not simply postulate the Dedekind infinity of *Principia*'s reconstructed numbers, but derived it as a nontrivial theorem from the assumption that the class of individuals is noninductive.[12] The difficulty with *Principia*'s use of the Axiom of Infinity is therefore not that it assumes what it sets out to prove; the difficulty is that the axiom is without any justification other than the fact – assuming it is a fact – that it happens to be true. As we will see, this undermines *Principia*'s account of arithmetic by leaving it open to the objection that even if its reconstruction of the numbers is based on true premises, it cannot provide a correct analysis of the epistemological basis for our belief that the numbers are Dedekind-infinite.

Frege avoided an appeal to a postulate like *Principia*'s by exploiting a subtlety in the logical form of the cardinality operator to argue that if our conception of a number is of something that can fall under a concept of first level, then the criterion of identity ensures that the numbers must form a Dedekind-infinite class. The idea that underlies Frege's derivation of the infinity of the numbers from his criterion of identity is the assumption that the cardinality operator is represented by a mapping from concepts to objects. In Frege's hierarchy of concepts and objects, the value of the mapping for a concept as argument has a logical type that allows it to fall under a concept of first level. This has the consequence that any model of Hume's principle must contain infinitely many objects. Beginning with the concept under which nothing falls, and whose number is by definition equal to zero, Frege is able to formulate the notion of a series of concepts of strictly increasing cardinality

---

[12] *Principia*, vol. II, *124.57. For a discussion of the relation between the notion of infinity assumed by *Principia*'s axiom and Dedekind infinity, as well as a discussion of Russell's proof – without the Axiom of Choice, but assuming the Axiom of Infinity – that the cardinal numbers of *Principia* form a Dedekind-infinite class, see Boolos (1994).

having the property that, with the exception of the first concept, every concept in the series is defined as the concept that holds of the numbers of the concepts which precede it. The idea of such a series of concepts is the essential step in Frege's derivation of PA^2 from FA. For Frege this derivation was a premise in a more general argument that was intended to illuminate the relation between arithmetic and logic. Frege had hoped to complete the argument by appealing to a demonstration of the existence of a domain of "logical objects" – classes or extensions of concepts – that could be identified with the numbers. His inability to establish the existence of such a domain has tended to obscure his achievement relative to later developments.

To be sure, Hume's principle, like *Principia*'s Axiom of Infinity, holds only in infinite domains. But it does not follow that the two accounts stand or fall together. A reconstruction of the numbers on the basis of the Axiom of Infinity is extraneous to arithmetic, both in its content and in its account of the epistemic basis of arithmetic. Frege, however, provided an account of the Dedekind infinity of the numbers in terms of their criterion of identity and the logical form of the cardinality operator. A central aspect of his account is lost in a reconstruction like *Principia*'s or one that treats numbers as higher-level concepts rather than objects. For, on any such account, the number of entities of any level above the type of individuals – and therefore also the number of numbers – depends on a parameter – namely, the cardinality of the class of possible arguments to first-level concepts – whose value can be freely set. But then on such a reconstruction the cardinality of the numbers can be freely specified.

By contrast, on Frege's account we can certainly consider domains of only finitely many arguments to first-level concepts – including domains consisting of a selection of only finitely many numbers. But a domain contains the numbers only if it satisfies their criterion of identity, and this forces the domain to be infinite, and the numbers Dedekind-infinite. Hence, whatever the technical interest of a reconstruction of the numbers like *Principia*'s or one that reconstructs the numbers as higher-level concepts, it does not share what is arguably the philosophically most interesting feature of Frege's theory of number: Frege's theory preserves the epistemological status of the pure theory of number by showing that the infinity of the numbers is a consequence of his analysis of number in terms of the criterion of identity.

1.4 ANALYSIS VERSUS JUSTIFICATION

A subtlety in the logical form of Hume's principle that we have emphasized makes it all the more compelling that the account of neither needing nor admitting of proof should not rest on a naive conception of self-evidence. The strength of Hume's principle derives from the fact that the cardinality operator is neither type-raising nor type-preserving, but maps a concept of whatever "level" to an *object*, which is to say, to a possible argument to a concept of *lowest* level. Were the operator not type-lowering in this sense, Frege's argument for the Dedekind infinity of the natural numbers would collapse. The fact that only the type-lowering form of the cardinality operator yields the correct principle argues against taking neither needing nor admitting of proof to be captured by self-evidence: it might be that only one of the weaker forms of the principle drives the conviction of "obviousness," "undeniability," or "virtual analyticity" that underlies what I am calling naive conceptions of self-evidence. Although it is highly plausible that the notion of equinumerosity implicit in our cardinality judgments is properly captured by the notion of one-to-one correspondence, a further investigation is needed to show that Hume's principle is "self-warranting," or whatever one takes to be the appropriate mark of neither needing nor admitting of proof.

Frege was at times highly sympathetic to the idea that fundamental principles can be justified on the basis of their sense. In *Function and Concept*[13] he endorsed the methodology of arguing from the grasp of the sense of a basic law to the recognition of its truth in connection with Basic Law v:

For any concepts *F* and *G*, the extension of the *F*s is the same as the extension of the *G*s if, and only if, all *F*s are *G*s and all *G*s are *F*s.

This lends plausibility to the relevant interpretative claim, but the fact that Law v arguably *does* capture the notion of a Fregean extension poses insurmountable difficulties in the way of accepting this methodology as part of a credible justification of it. Being analytic of the notion of a Fregean extension does not show Basic Law v to be analytic, true, or even consistent. If therefore it is a mark of primitive truths that our grasp of them suffices for the recognition of their truth, then some at least of Frege's basic laws are not primitive truths. Although grasping the sense of a basic law does not always suffice for the recognition of its truth, Frege never appears to have had a more considered methodology for showing that we are justified in believing his basic laws. He seems to have taken for granted that the basic laws of logic and arithmetic are self-warranting and that this is an assumption to which all parties to the discussion are entitled.

I have been concerned to show that Frege's emphasis on the generality of the premises employed in a proof of apriority – rather than their

---

[13] Frege (1891, p. 11 of the original publication).

independence from experience – is sustainable independently of the truth of logicism. Although Frege's analysis would preserve the notion that the principles of arithmetic express a body of truths that are known independently of experience, there is a respect in which it is independent even of the weaker claim that Hume's principle is a known truth. For, even if the traditional conception of arithmetic as a body of known truths were to be rejected, it would still be possible to argue that Hume's principle expresses the condition on which our application of the numbers rests. As such, it captures that feature that makes the numbers "necessities of thought" and gives our conception of them the constitutive role it occupies in our conceptual framework.

If we set to one side the question of the truth of Hume's principle, there is a fact on which to base the claim that it is foundationally secure, albeit in a weaker sense than is demanded by either the traditional or the Fregean notion of apriority. By the converse to Frege's theorem FA is recoverable from a definitional extension of PA^2, or equivalently, FA is interpretable in PA^2. As a consequence, FA is *consistent relative to* PA^2; a contradiction is derivable in FA only if it is derivable in PA^2, a fact that follows from Peter Geach's observation[14] that when the cardinality operator is understood as a type-lowering mapping from concepts to objects, the ordinals in $\omega + 1$ form the domain of a model of FA. Since Frege regards the basic laws of arithmetic to be known truths, the interpretability of FA in PA^2 would certainly count as showing that FA is foundationally secure as well. But Frege would likely have regarded such an argument as superfluous since Hume's principle was for him also a known truth, and therefore certainly foundationally secure. This shows why an appeal to the *consistency strength* of FA in support of Frege's foundational program must be judged altogether differently from its use in the program which motivated the concept's introduction into the foundations of mathematics.

The study of the consistency strength of subtheories of PA^2 is an essential component of the program we associate with Hilbert, namely, to establish the consistency of higher mathematics within a suitably restricted "intuitive" or "finitistic" mathematical theory. Gödel's discovery of the unprovability of the consistency of first-order Peano arithmetic (PA) within PA (by representing the proof of the incompleteness of PA within PA) motivated the investigation of subtheories of PA, incapable of proving their own consistency, and extensions of PA, capable of proving the consistency of PA. The theory Q known as Robinson arithmetic is a particularly simple

---

[14] In Geach (1976). See Boolos (1987) for a full discussion.

example of a theory incapable of proving its own consistency, and it forms the base of a hierarchy of increasingly stronger arithmetical theories.[15] But the study of this hierarchy is not integral to the foundational program of Frege for whom knowledge of the truth, and therefore the consistency of PA^2 (and thus of PA), is simply taken for granted.

Of the two foundational programs, only Hilbert's holds out any promise of providing a foundation which carries with it any real justificatory force. Frege's foundational focus differs from Hilbert's in precisely this respect. The goals of Frege's logicism are epistemological, but they are not those of making the basic laws of arithmetic more secure by displaying their basis in Hume's principle, or indeed in logic. To use our earlier terminology, Frege's derivation of PA^2 from FA is part of an *analysis* of arithmetical knowledge rather than a *demonstration* of it.[16]

Indeed, Hilbert's proposal for securing PA^2 and its set-theoretical extensions on an intuitive basis runs directly counter to Frege's goal of showing arithmetic to be analytic and hence independent of intuition. This is because Hilbert's view depends on taking as primitive the idea of iteration (primitive recursion) and our intuition of sequences of symbols:

[T]hat the [symbolic objects] occur, that they differ from one another, and that they follow each other, or are concatenated, is immediately given intuitively, together with the object, as something that neither is reducible to something else nor requires such reduction. (Hilbert 1925, p. 376)

An attractive feature of the program of basing the consistency of PA^2 on some particularly weak fragment of arithmetic is that it held out the promise of explaining *why* the theory on which it proposed to base the consistency of

---

[15] As Saul Kripke has observed, since in Q one cannot even prove that $x \neq x + 1$, a natural model for Q is the cardinal numbers with the successor of a cardinal x defined as $x + 1$. Kripke's observation is reported in Burgess (2005, p. 56). Burgess's book describes where a variety of subtheories and extensions of FA are situated in relation to this hierarchy; it should be consulted by anyone interested in the current state of art for such results.

[16] See John Bell (1999b) for a formulation of the relationship between certain distinguished models of Hume's principle – "Frege structures" – and models of the standard set-theoretic reconstruction of Peano–Dedekind arithmetic ("Peano–Dedekind models"). Bell proves the existence of a procedure f which converts a minimal Frege structure into a Peano–Dedekind model and a procedure g which converts a Peano–Dedekind model into a minimal Frege structure; he then shows that f and g are mutually inverse to one another. This has as a consequence the equivalence of the existence of Peano–Dedekind models and the existence of minimal Frege structures. The importance of this fact is not foundationalist in the sense that it shows one or another conception to be a more secure basis for our knowledge of number; such an interpretation is evidently excluded by the interchangeability of the two representations. Rather, the fact that f and g are mutually inverse to one another shows that the notions of a Peano–Dedekind model and a minimal Frege structure provide two different but thoroughly interchangeable representations of our informal conception of number, one associated with the pure theory of number, the other with its application in judgments of cardinality.

PA^2 is plausibly viewed as self-warranting. The notions which such a theory permits us to express – the intuitability of finite sequences of symbols and the iterability of basic operations involving them – are arguably presupposed in the formulation of *any* theory. But a theory that rests on no more and no less than what is demanded by the formulation of a mathematical theory is clearly as warranted as any such theory.

Although Frege's *Grundgesetze* contains a theory of general principles by which to explain the apriority of our knowledge of arithmetic, it requires an appeal to arithmetic's basis in logic. Rather surprisingly – and independently of the issue of consistency – the nature and intended scope of the system of logic of *Grundgesetze* prevent it from possessing the simple intuitive appeal of Hilbert's proposal. Were it not for Gödel's second incompleteness theorem, Hilbert's approach to the consistency of PA^2 would have constituted a successful transcendental justification of it, since it would have shown that PA^2 is consistent relative to the framework of assumptions that underlie the possibility of formulating a mathematical theory. Hilbert therefore promised to carry the argument for the apriority of arithmetic in a direction that is genuinely minimalist in its use of primitive assumptions. This cannot be said of the system of *Grundgesetze*. For this reason, Hilbert's account is more recognizably a traditional foundation or justification for arithmetic – indeed, for mathematics generally – than Frege's.

Our emphasis has been on Frege's analysis of those basic laws of arithmetic which ensure the Dedekind infinity of the numbers. Early on, Hilbert expressed concern with the method by which the existence of a Dedekind-infinite concept is proved in the logicist tradition.[17] But the principal divergence between logicist and nonlogicist approaches to arithmetic arises in connection with the explanation of the remaining law – the principle of mathematical induction. Frege explains the validity of reasoning by induction by deriving the principle from the definition of the natural numbers as the class of all objects having all hereditary properties of zero. Anti-logicists explicitly reject this explanation, since it rests on the questionable idea of the totality of all properties of numbers; they argue that in light of the paradoxes our confidence in this notion is not justified.

Although it was conceived in ignorance of the set-theoretic paradoxes and the analysis of their possible source, there is a sense in which the program of recovering arithmetic from FA retains its interest and integrity even in light of the paradoxes: Frege's definition of the natural numbers shows that there is also a *basis* – if not an entirely satisfactory *justification* – for

---

[17] See especially the paragraphs devoted to Frege and Dedekind in Hilbert (1904, pp. 130–1).

characteristically arithmetical modes of reasoning in general reasoning, and therefore in conceptual thought itself. It is a further discovery that pursuing arithmetic from this perspective is susceptible to a doubt that might appear avoidable if we take as primitive the idea of indefinite iteration. But it remains the case that there is a compelling account of reasoning by induction that recovers it from general reasoning about concepts, even if our confidence in such an account is diminished by various analyses of the paradoxes.[18]

1.5 SUMMARY AND CONCLUSION

Frege extracted from his reflections on Kant several desiderata that an account of number should meet for it to be independent of intuition. First, such an account must explain the *applicability* of numerical concepts without invoking any intuitions beyond those that are demanded by the concepts to which the numerical concepts are applied. Secondly, it must explain our understanding of recognition judgments without recourse to intuition, and in terms of relations among concepts. Thirdly, the account must recover any piece of arithmetical reasoning that rests on the possibility of indefinitely iterating an operation as a species of general reasoning. Frege tells us (*Gg*, p. ix) that the "fundamental thought" on which his analysis of the natural or counting numbers is based is the observation (formulated in sect. 46 of *Grundlagen*) that a statement of number involves the predication of a concept of another concept; numerical concepts are concepts of second level, which is to say, concepts under which concepts of first level – concepts which hold of objects – are said to fall. This yields an analysis of the notion of a numerical property, as when we predicate of the concept *horse which draws the King's carriage* the property of having four objects falling under it.

To pass from the analysis of numerical properties to the numbers, Frege introduced the cardinality mapping – *the number of Fs*, for *F* a sortal concept – which is "implicitly defined" by the *criterion of identity*:

The number of *Fs* is the same as the number of *Gs* if and only if the *Fs* and the *Gs* are in one-to-one correspondence.

This criterion is the basic principle upon which Frege's development of the theory of the natural numbers is based: it states the condition under which judgments to the effect that the number of *Fs* is the same as the number of *Gs* are true.

---

[18] Particularly Russell's; see Chapter 10 below for a discussion of his analysis of the paradoxes.

The intuitive idea underlying Frege's derivation of the Dedekind infinity of the numbers is the definition of an appropriate sequence of representative concepts. The existence of a number for each of the concepts in the sequence is established using only logically definable concepts and those objects whose existence and distinctness from one another can be proved on the basis of the criterion of identity. Nowhere in this construction is it necessary to appeal to extensions of concepts. In its mathematical development Frege's analysis thus provides an account of our understanding of the finite cardinal numbers and the first infinite cardinal, one which can be carried out independently of the portion of his system which led to inconsistency, and one which is demonstrably consistent.

Frege's earliest contribution to the articulation of logicism consisted in showing that the validity of reasoning by induction can be explained on the basis of our knowledge of principles of reasoning that hold in *every* domain of inquiry. This directly engages Frege's understanding of the Kantian tradition in the philosophy of the exact sciences, according to which principles of general reasoning peculiar to our understanding must be supplemented by intuition for an adequate theory of arithmetical knowledge. In *Begriffsschrift* Part III Frege showed how the property of, for example, following the number zero in the sequence of natural numbers after some finite number of relative products of successor might be defined without reference to number or finiteness. The principle of mathematical induction is an almost immediate consequence of this definition.

Frege's explanation of the possibility of arithmetical knowledge relies on higher-order features of his logic. The assessment of Frege's success in addressing the basis for our knowledge of the infinity of the numbers and the justification of induction, by contrast with the assessment of his success in connection with the applicability of arithmetic, has stalled on the question whether second-order logic is really logic. This is unfortunate, since it has deflected attention from Frege's claim to have shown that reasoning by induction and our knowledge of a Dedekind-infinite system depend on principles which, whether they are counted as logical or not, enjoy the generality and universality of conceptual thought; they do not therefore rest on a notion of Kantian intuition of the sort Frege sought to refute. As Frege understood it, such a notion would require intuition of objects, and it is precisely a reliance on the intuition of objects that he avoided. It may be necessary to concede that Frege's demonstrations rest on intuition in a familiar, psychologistic, sense of the term; what has been missed is how little is conceded by such an admission.

CHAPTER 2

Carnap's thesis

By *Carnap's thesis* I mean the assertion that certain applied mathematical theories are not *factual*. This thesis does not stake out a position on the metaphysical question of the reality of mathematical objects, or more generally on realism in the philosophy of mathematics. Nor is it merely an extension of Carnap's well-known "principle of tolerance." Carnap's thesis is a distinct and independently defensible claim that seeks to illuminate the methodological status of different applied mathematical theories.

My proposal for how Carnap's thesis should be understood makes essential use of Frege's notion of a *criterion of identity*. I will explain how Frege's notion can form the basis of a general framework for addressing the distinction between applied mathematical theories with and without a factual content. For example, it has long been held that the distinction between factual and nonfactual applied mathematical theories is well illustrated by geometry and arithmetic. We will see how the distinction between these two cases arises from a difference in the criteria of identity that are appropriate to their application – the criterion of identity that controls ordinary applications of the Peano–Dedekind theory of the natural numbers in the case of arithmetic, and the criteria of identity that control the standard applications of the theories of Euclid and Minkowski in the case of geometry. To anticipate what will be elaborated in detail below, the factuality of an applied mathematical theory will be shown to turn on the recognition that the criteria of identity appropriate to its objects are *empirically constrained* in a way that the criteria of identity appropriate to the objects of a nonfactual theory are not. A novel aspect of this elaboration and defense of Carnap's thesis is how the thesis emerges as a more nuanced and preferable alternative to those accounts of the distinctive place of arithmetic and analysis in our conceptual framework that are based on Quine's notion of centrality.

2.1 BACKGROUND TO THE THESIS

Frege and Carnap are both identified with the idea that arithmetic, in the broad sense which includes the theory of the real numbers, is analytic. But there were considerable differences between them, both in how they understood this claim and in how they thought it could be established. These differences affected their respective views of the philosophical interest that would attach to a demonstration of the analyticity of arithmetic. Frege promised a simple and elegant but also very partial answer to Kant's question, 'How is synthetic a priori knowledge possible?' By showing that arithmetic can be recovered as a part of logic, an important candidate for synthetic a priori knowledge would be revealed as nothing more than logical knowledge. Thus, for Frege, the interest of a demonstration of the analyticity of arithmetic consisted in what it would show about the relation of arithmetic to logic and the consequences of this for the scope of Kant's notion of a priori intuition. But as the dialectic unfolded it became clear that unless one is prepared to count simple type theory with an Axiom of Infinity as logic, it is not possible to derive arithmetic from logic. And if Carnap ever did endorse such a liberal conception of logic it was not in order to reduce arithmetic to it, since by the time of *Logical Syntax of Language*,[1] we find that the list of axioms or "primitive sentences" of Language II of that work already includes the Peano–Dedekind axioms.

Carnap agreed with Frege that arithmetical reasoning is captured by logical reasoning. But the fact that Carnap did not pursue a reduction of arithmetic to logic – even while continuing to maintain its analyticity – suggests that his understanding of analyticity and his conception of the problem of synthetic a priori knowledge were very different from Frege's. And indeed Carnap's target in the debate over the synthetic a priori is not so much Kantian intuition as it is the rationalist thesis that reality is capable of being grasped by pure thought alone. By contrast, Frege never addressed the synthetic a priori in this form. Not only did Carnap have a much broader conception of the problem of synthetic a priori knowledge, his interest in the problem had an empiricist motivation that is totally alien to Frege. So far as arithmetic is concerned, the central claim that Carnap was committed to defending was not the proposition that arithmetic is a part of logic, but the thesis that, like logic, it is not "factual" in a sense that I will be concerned to elucidate and clarify.

---

[1] Carnap (1934); cited as *LSL*.

Although my goals are not historical, there are some aspects of Carnap's thought that the following discussion does seek to clarify. Among these I should emphasize two points of philosophical method, one concerning the bearing of Carnap's thesis on the problem of the synthetic a priori, and the other addressing the sense in which Carnap advanced his thesis as a philosophically significant claim. However, my principal goal is not the interpretation of Carnap, but the presentation of a sympathetic reconstruction and defense of his thesis, focusing especially on its application to arithmetic and geometry. Unlike an interpretation, a reconstruction is free to depart from the actual historical development if a more satisfactory alternative suggests itself. One such departure is my account of the application of arithmetic, where I argue that despite the fact that Carnap's thesis is not one Frege sought to defend, his views on applicability support and illuminate it in a surprising way. Carnap always stressed the importance of Frege's emphasis on applicability for a complete account of the nature of mathematics.[2] But my use of Frege's contributions in support of Carnap's thesis has no precedent in Carnap's writings.

2.2 TWO STRATEGIES

Carnap's commitment to the nonfactuality of logic and mathematics derived from a variety of sources. He himself invariably cited Wittgenstein's *Tractatus*[3] as his principal inspiration. Without going into the details of Wittgenstein's doctrine that the propositions of logic are *sinnlos*, it is safe to say that for Carnap the essential point of Wittgenstein's analysis is that unlike genuine propositions, tautologies and contradictions fail to partition states of affairs into those that obtain and those that fail to obtain. But according to the *Tractatus* a proposition is informative only to the extent that it makes such a partition. The fact that logical propositions fail to be informative in this sense is what underlies the Tractarian idea that they are without factual content: *failing to partition states of affairs, they assert no facts.*

Carnap was well aware that Wittgenstein's analysis of logical truth in terms of tautologousness is severely limited and that it needs to be supplemented if it is to cover interesting cases that arise in "higher logic" and mathematics. The use of tautologies even at this illustrative stage of the

---

[2] See especially *LSL* (sect. 84). For a discussion of Frege's influence on Carnap, see Reck (2004) and (2007).
[3] Wittgenstein (1922).

dialectic suggests that Carnap's strategy for establishing the nonfactuality of mathematics is to replace tautologousness by analyticity ("A-truth") and to then argue from the nature of analyticity to the nonfactuality of propositions that are classified as A-true. Such a strategy assumes that the ease with which one can defend the nonfactuality of tautologies will carry over to an argument from the broader notion of analyticity. I call this strategy for securing the nonfactuality of logic and mathematics *the Tractarian strategy*. The characterization of this strategy as Tractarian is not intended as an interpretive claim regarding the *Tractatus*, since that work treats logic and mathematics very differently, but as a suggestive allusion to the Tractarian influence on Carnap's articulation of his thesis.[4]

Here is a capsule summary of the Tractarian strategy's motivation and principal goal: had the propositions required for the reduction of mathematics to logic been mere tautologies, there would be no difficulty maintaining the claim that mathematical propositions are nonfactual. But since logicism in this form is unworkable, it is necessary to broaden the notion of analyticity beyond mere tautologousness and argue that any proposition which is analytic is as evidently nonfactual as a simple tautology. The view that analyticity can be explained as truth in virtue of meaning is one possible development of the Tractarian strategy, since what is true in virtue of

---

[4] In his *Intellectual Autobiography*, Carnap says of Wittgenstein:

> The most important insight I gained from his work was the conception that the truth of logical statements is based only on their logical structure and on the meaning of terms. Logical statements are true under all conceivable circumstances; thus their truth is independent of the contingent facts of the world. On the other hand, it follows that these statements do not say anything about the world and thus have no factual content. (1963a, p. 25)

My remarks in the text extrapolate from the statements of *logic* to the statements of *logic and mathematics*. Carnap endorses this extrapolation a little later in the *Autobiography*, when he recalls that:

> to the members of the Circle there did not seem to be a fundamental difference between elementary logic and higher logic, including mathematics. Thus we arrived at the conception that all valid statements of mathematics are analytic in the specific sense that they hold in all possible cases and therefore do not have any factual content. (1963a, p. 47)

A similar view can be found as early as *LSL*, where it is combined with an explicit rejection of Frege's program of deriving mathematics from logic:

> Whether … only logical symbols in the narrower sense are to be included amongst the primitive symbols … or also mathematical symbols … and whether only logical primitive sentences … or also mathematical sentences, is not a question of philosophical significance, but only one of technical expedience. (*LSL*, p. 327)

It should also be noted that Carnap here distinguishes between questions of philosophical significance and questions of technical expedience. It seems safe to assume that Carnap is not being ironic and that the distinction to which he is appealing is not between two classes of questions, one of which is the empty class.

meaning is not informative in the intended sense, and therefore not true in virtue of the relevant kind of fact.

Although this is not necessarily the only way of developing the Tractarian strategy, the idea that the defense of Carnap's thesis demands some form of the strategy is very widespread. Certainly Quine's "Two dogmas"[5] can be understood in this way. As is well known, Quine argues that the main obstacle to the success of Carnap's program is the difficulty involved in extending the notion of analyticity beyond that of tautologousness and first-order logical truth to a notion that can fulfill the philosophical aims for which it was intended. The problem, according to Quine, is that to get a suitably broad notion of analyticity Carnap's proponents have either appealed to an inherently vague notion of meaning, or have simply stipulated that certain statements are analytic. Quine argued that we can be confident of the analyticity of a stipulation like a definition (that meets recognized standards of correctness) but we can have no confidence that the propositions of mathematics are without factual content just because they have been stipulated to be analytic.

The difficulty observed in sect. 4 of "Two dogmas" (the section entitled "Semantical rules") is that merely stipulating the analyticity of the propositions of logic and mathematics cannot support the philosophical claims which their characterization as analytic is supposed to warrant. Quine concluded that the only viable candidate for explicating analytic truth is one based on a proposition's centrality to our conceptual framework. But not only is centrality not necessarily confined to the propositions of logic and mathematics, it is also a property that may be withheld from them. Hence, so far as analyticity and factuality are concerned, logic and mathematics must be treated on a par with the rest of the theoretical edifice of science. Quine's introduction of centrality thus threatens to undermine any interesting application of the factual–nonfactual distinction to different components of our conceptual framework. This is the key problem Quine's proposed reduction of analyticity and apriority to centrality poses for Carnap's thesis, and it is a topic to which I will return.[6]

---

[5] Quine (1951a).

[6] It is unclear whether Quine ever understood Carnap's thesis as a claim that is detachable from some form of conventionalism. Despite its well-known and often-quoted final paragraph alluding to the complex intertwining of fact and convention in the lore of our fathers, even "Carnap and logical truth" (Quine 1963) treats the nonfactuality of logic only as a corollary to one or another species of conventionalism. For Quine, the central question posed by logic is not whether it is factual, but how to account for its certainty. Mathematics enters the discussion through the paradoxes of set theory, since they serve to temper the enthusiasm with which a solution to the problem of logical certainty might be extended to a broader class of truths.

With the notable exception of Goldfarb and Ricketts (1992) most of Carnap's interpreters have regarded Quine's criticisms as decisive, and even Goldfarb and Rickets concede Quine's criticisms for any interpretation of Carnap that takes him to be making a substantive philosophical claim.[7] Their answer to Quine on Carnap's behalf is that Carnap does not claim to *know* that logic and mathematics are nonfactual. On their view Quine has simply failed to grasp the extent of Carnap's philosophical radicalism: Carnap is not a mere critic of the synthetic a priori. He rejects a funda-mental presupposition of the debate surrounding the concept. Goldfarb and Ricketts argue that Carnap is not advancing philosophical arguments on behalf of a thesis but is merely making a proposal about what propositions are analytic or nonfactual: Carnap proposes that the class of analytic state-ments should include the whole of logic and mathematics. On their interpretation, Carnap is not concerned to *warrant* or *enhance our confidence in* the thesis that logic and mathematics are nonfactual. Indeed, this brings us to what, according to Goldfarb and Ricketts, Carnap sees as the mistaken presupposition of the debate over the synthetic a priori: namely the assumption that it is concerned with theses which are correct or incorrect, warranted or not, when it can only be a matter of practical decisions which are better or worse. According to Goldfarb and Ricketts, this "better or worse" is to be understood relative to the goal of presenting the language of science in a form that reflects a freely adopted philosophical viewpoint: there is no question of showing the correctness or otherwise of such a viewpoint.

I hope to show that there is an alternative strategy for understanding and establishing Carnap's thesis. On this alternative strategy, the thesis emerges as an illuminating methodological analysis of the application of mathemat-ical theories in science. So understood, the thesis does not depend on *truth in virtue of meaning* or *truth by stipulation*. It is therefore not susceptible to the difficulties Quine found with those notions, and it owes little or nothing to the *Tractatus*. According to this approach to the justification of Carnap's thesis, the special status the thesis accords certain applied mathematical theories within our conceptual framework is not tied to the success of logicism, but derives from reflection on the epistemic status of geometry in connection with the theory of space and time.

The epistemological significance of the separation of pure from applied mathematics was made famous by Einstein in his Prussian Academy lecture

---

[7] See also Goldfarb (1997) and Ricketts (2007). Friedman (2006) is an exception; it advances an interpretation of Carnap that is much closer to the view presented here; see footnote 11, below.

of 1921[8] where he explained the distinction in terms of the difference between cases in which geometry does, and cases in which it does not, "refer to reality":

> In so far as the mathematical theorems refer to reality, they are not certain, and in so far as they are certain, they do not refer to reality ... The progress entailed by axiomatics consists in the sharp separation of the logical form and the realistic and intuitive contents ... The axioms are voluntary creations of the human mind ... To this interpretation of geometry I attach great importance because if I had not been acquainted with it, I would never have been able to develop the theory of relativity.[9]

Einstein's distinction has gone largely unchallenged for its clarification of the role of geometry in physics. It explains how, prior to the discovery of non-Euclidean geometry and the constructive proofs of its consistency relative to Euclidean geometry, there had been a persistent failure to separate geometry as a piece of pure mathematics that could be grasped a priori from geometry as a part of physics. As a result of this unclarity, geometry was thought to exemplify the possibility that physical reality can be known by pure thought alone.

It is true that important issues remained in connection with the role of convention in the interpretation of geometry as a component of our theory of physical space. Conventionalists, Carnap among them, sought to undermine the factuality of geometry in a stronger sense, arguing that some empiricist conclusions based on the separation of pure from applied mathematics oversimplified the methodological issues that are raised by the application of geometry in physics, so that even in the case of a theory of the geometry of physical space, there was some controversy involving the factual character of certain of its propositions. However this should not distract us from the fact that such conventionalist considerations are – and were seen to be by the conventionalists themselves – entirely compatible with the essential soundness of Einstein's assessment that the application of geometry rests on a number of evidently empirical and unproblematically factual assumptions.

I will call the approach to Carnap's thesis that has its source in Einstein's elaboration of how a geometrical theory is applied *the Einsteinian strategy*.[10] According to this strategy, Carnap's thesis is an informative claim about the methodological character of applied mathematical theories. It is not informative because it has true empirical consequences. It may have such

---

[8] Einstein (1921); translated and published as "Geometry and experience."
[9] Quoted by Freudenthal (1962, p. 619).   [10] Thanks to Kai Wehmeier for suggesting this term.

consequences regarding our use of 'factual' and 'nonfactual,' but if it does, this is irrelevant to the intended significance of the claim that the thesis is an informative one. Rather, the thesis is informative because it clarifies a number of conceptual issues that had become entangled in traditional philosophical and scientific discussions of the nature of geometry; and as we will also see, it is informative because of the light it casts on the difference between arithmetic and geometry.[11]

There is a clear difference between the Tractarian and Einsteinian strategies. The Tractarian strategy aims to derive nonfactuality from the concept of analyticity as, for example, the development of the strategy in terms of explications of analyticity as tautologousness, truth in virtue of meaning, and truth by stipulation proposes to do. By contrast, the Einsteinian strategy proceeds by showing how, on grounds that are independent of the analysis of analyticity, pure geometry is nonfactual, while applied geometry is factual. Thus, the Einsteinian strategy does not base its account of geometry on an appeal to analyticity, still less on the notion of meaning, but on the success of a methodological analysis which proved fundamental for the light it cast on the relation between pure and applied mathematical theories and the physics of space and time.

But as attractive as the Einsteinian strategy is in connection with the distinction between pure and applied *geometry*, it is far from obvious how it should be used to clarify the status of applied *arithmetic*. The importance of distinguishing geometry from arithmetic is emphasized in Carnap (1939) where "Geometry and experience" is quoted at length. Carnap recognizes that the distinction must be drawn at the level of *applied* arithmetic and *applied* geometry, because "[t]he doctrine [that arithmetic and geometry are synthetic a priori] was obviously meant to apply to arithmetic and geometry as theories, i.e. as interpreted systems, with their customary interpretations." He then argues that since the customary interpretation of arithmetic is a "logical" rather than a "descriptive" one, arithmetic is "independent of experience [and does not] concern experience or facts at all . . . On the basis of the customary interpretation, however, the sentences of geometry, as propositions of physical geometry, are indeed factual (synthetic), but dependent upon experience, empirical."[12] In my view, Carnap's proposal

---

[11] This is entirely consonant with Friedman's emphasis on Carnap's "repeatedly expressed conviction, characteristic of his semantical period, that the distinction between analytic and synthetic truth 'is indispensable for the logical analysis of science,' so that 'without [it] a satisfactory methodological analysis of science is not possible.'" See Friedman (2006, p. 50) and the citations there of Carnap (1942, p. xi; 1963b, p. 932; and 1966, p. 257).

[12] All quotations in this paragraph are from Carnap (1939, p. 198).

for drawing the factual–nonfactual distinction on the basis of whether the vocabulary of a theory is descriptive or logical runs the risk of reducing the distinction to one of conventional choice rather than methodological principle. In order to address these issues it is necessary to look in greater detail at how Einstein illuminated our understanding of the application of a geometrical theory. We can then turn to the application of arithmetic and the consideration of how it differs from geometry.

2.3 APPLYING GEOMETRY

It is natural to look to Einstein's "Geometry and experience" for his account of the application of a geometrical theory. The paper focuses on the question: How can Euclidean geometry, which Einstein refers to as "the most ancient branch of physics," be a natural science when it is concerned with the disposition of bodies – perfectly rigid bodies – whose existence is precluded by physics? Einstein's objective is to counter Poincaré's view that since there are no perfectly rigid bodies, the choice of geometry is conventional, and may therefore be guided by a "principle of simplicity." According to Poincaré, such a principle will always favor Euclidean geometry. Einstein's answer to Poincaré is based on his introduction and defense of the notion of the "practically rigid body" and his correlative formulation of the principle of free mobility in terms of this notion. Einstein's discussion of these matters is of great importance, and it will occupy our attention in a moment. But as instructive as this discussion is for our understanding of how Euclidean geometry can be represented as a natural science, the methodological contribution that will prove to be of particular interest for us is the discussion of simultaneity in Einstein (1905). Although the explicit introduction of Minkowski geometry post-dates the 1905 paper, Einstein's analysis of simultaneity implicitly contains a general framework for adjudicating among theories of space-time structure and an explanation of how a four-dimensional geometrical theory like Minkowski's is applied.

There are strong reasons for placing special emphasis on Einstein's contribution to our understanding of the application of Minkowski geometry rather than his discussion of the nature and status of the claim that space is Euclidean: (1) The application of Minkowski geometry raises a number of issues, but the "ideal" character of geometrical objects is far from being the most important among them; bypassing this question makes it possible to set the contrast between geometry and arithmetic in sharp relief. (2) The problem Einstein addressed in 1905 was conceptually transformative in a way that the problem posed by Euclidean geometry is not: Einstein

(1905) articulated a framework without which it was completely unclear how a question about the relativity of time could even be formulated, let alone answered. (3) As we will see, the issues of applicability that Minkowski geometry raises are in an important sense *prior* to the question of conventionalism, and, in fact, the account that can be extracted from Einstein's analysis of time and simultaneity shows that the choice cannot be a wholly conventional one.

Einstein's investigations into the role of space and time in physics began with an analysis of the concept of an inertial frame and the observation that its understanding requires a closer consideration of the concept of simultaneity. An inertial frame associates a spatial coordinate system, and therefore a notion of distance, with a free motion. But the relevant notion of distance is evidently one that relates distances *at a time*; it therefore presupposes the concept of simultaneity. While it might seem natural to make the "absolutist" assumption that all inertial frames employ the *same* relation of simultaneity, the very fact that we can question this assumption shows that it is not something that should be pre-judged, as it would be if the assumption of the absoluteness of simultaneity were built into the notion of an inertial frame, as it is in classical mechanics.

What should we demand of a framework for investigating the question of the absoluteness of simultaneity? A natural first step is to ask what our judgments that distant events occur at the same time rest upon. Einstein noted that even in the simplest cases, as when two events are seen to be simultaneous, there is a tacit assumption involving the transmission of light between their respective places of occurrence. He then extrapolated from this observation to the idea that any physical criterion for recognizing the time of occurrence of two distant events as the same must accord with one that is based on the transmission of a signal. If we were in possession of a signal whose speed is constant and independent of the speed of the source of its emission, we would have a procedure for establishing when two events occur at the same time. Such a procedure would not depend on the inertial frame which deploys it and would settle whether simultaneity is the same in all inertial frames without assuming that it is from the outset. By Maxwell's theory, light is such a signal. Einstein therefore proposed that we establish the simultaneity of distant events – i.e. their simultaneity relative to an inertial frame – by employing light signals.

Given this framework, the relevant part of Einstein's analysis may be explained as follows. Figure 2.1 depicts two free particles in relative motion to one another whose paths A and B intersect in the event e. According to Einstein's signaling procedure, the time of occurrence of a distant event is

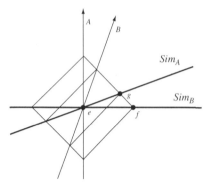

Figure 2.1

the same as the time of occurrence of *e* if *e* bisects the interval along *A* (*B*) that is bounded by the events marking the emission and reception of a light signal sent from a point on *A* (*B*) and reflected back to *A* (*B*) from this distant event. The fact that the velocity of light is the same for an inertial frame defined with respect to *A* as for one defined with respect to *B* implies that the class of events simultaneous with *e* will differ for different inertial frames. In a two-dimensional space-time, the lines of simultaneity *Sim$_A$* and *Sim$_B$* for the inertial frames associated with *A* and *B* are just the ("space-like") diagonals of the appropriate "light parallelograms" centered on *e*. And according to this analysis, the time of occurrence of *e* is the same as the time of occurrence of a distant event *f* (*g*) just in case *e* and *f* (*g*) belong to a common simultaneity line constructed on the basis of such a configuration of emitted and reflected signals.

It will be noticed that Einstein's analysis turns on the provision of a *criterion for identifying* the times of occurrence of spatially separated events. If, on the basis of experimental considerations, we reject the assumption of an ether frame, the relation of simultaneity to which we are led by Einstein's analysis is fundamentally different from the absolute relation assumed by Newtonian mechanics, a fact that is displayed in the figure by the multiplicity of lines of simultaneity through *e*.[13] The

---

[13] Since the Newtonian framework is antecedently committed to the absoluteness of simultaneity, and since the theory of gravitation is a derived theory within Newtonian mechanics, even if Newtonian gravitation could be made to yield a practical procedure for determining the simultaneity of distant events, the result of applying such a procedure would necessarily have to conform to those spatio-temporal assumptions of the Newtonian framework that are reflected in the classical addition law for velocity. Thus, without Einstein's analysis, the a posteriori character of our knowledge of simultaneity

application of a four-dimensional geometrical theory is thus transformed into a part of physics by virtue of the fact that Einstein's criterion of identity for *time of occurrence* invokes nomological properties of the signaling relation by which events are connectible.

In "Geometry and experience" Einstein provides an analysis of the basic equivalence relation of geometry – namely, congruence – that is similar to his analysis of simultaneity. Here the role of the constancy of the speed of light and its independence of the state of motion of its source is played by Einstein's formulation of the principle of free mobility in terms of practically rigid bodies. The formulation arises in the context of Einstein's discussion of a notion for which the standard English translation of "Geometry and experience" uses the not particularly useful word 'tract.' But if we think of a tract as a bounded line-segment on a practically rigid body, then in his analysis of free mobility, Einstein effectively formulates the following *criterion of identity for the lengths of tracts*:

The length of one tract is the same as the length of another if and only if the tracts are congruent,

where *congruence* is constrained by the principle of free mobility:

If two tracts are found to be congruent once and anywhere, they are congruent always and everywhere.[14]

But since this principle is established experimentally, the criterion of identity for the lengths of tracts is empirically constrained, and applied geometry is factual.

---

would have been missed. This is not to say that the Newtonian framework is incoherent because simultaneity is only theoretically definable in it. The difficulty is that such a theoretically defined relation cannot be the limiting case of a procedure which exploits less than infinitely fast signals should such a procedure turn out to be incompatible with the absoluteness of simultaneity. But this is precisely the situation with light signaling: light's velocity is not only greatest among the velocities of signals that are capable of serving as practical measures of simultaneity. Experiments also show that, contrary to the Newtonian velocity addition law, the velocity of light is the same for all inertial frames. The invariance of the velocity of light therefore undermines the assumption that a definition of simultaneity in terms of light signals can be understood as part of a process of approximation to the theoretically defined relation of absolute simultaneity that is assumed by the Newtonian framework. See DiSalle (2006, Chapter 4, sect. 2) where these points are taken up in some detail.

[14] The full text of the relevant passage from Einstein (1921, p. 37) reads:

We will call that which is enclosed between two boundaries, marked upon a practically-rigid body, a tract [*Strecke*]. We imagine two practically-rigid bodies, each with a tract marked out on it. Two tracts are said to be 'equal to one another' if the boundaries of the one tract can be brought to coincide permanently with the boundaries of the other. We now assume that . . . [i]f two tracts are found to be equal once and anywhere, they are equal always and everywhere.

2.4 APPLYING ARITHMETIC

Frege's development in *Grundlagen*[15] of his positive view of the applicability of arithmetic is restricted to the simplest case, that of the application of the theory of the natural numbers in counting. Frege begins with an explanation of the content of a statement of number as an assertion about a concept. After a brief but significant digression motivating the view that numbers are objects rather than concepts, Frege continues with a discussion of *recognition judgments* – judgments expressed by sentences of the form,

The number of *F*s is the same as the number of *G*s,

where *F* and *G* are any sortal concepts whatsoever. Of course, the cases that will be particularly relevant for us are those in which *F* and *G* are empirical concepts; but Frege's discussion of applicability is not in any way restricted to such cases.

The interest of recognition judgments for Frege consists in the fact that because they express the judgment that the same number is "given to us" in two different ways, as the cardinal number of two different concepts, they are of fundamental importance to the defense of his thesis that numbers are "self-subsistent objects." But in the course of defending this admittedly metaphysical thesis, Frege is led to the concept that forms the foundation of his account of the application of arithmetic when he asks for the criterion by which we assess the truth of a recognition judgment. Frege's answer consists in his *criterion of identity for number*: a recognition judgment involving numbers is true just in case the concepts it concerns are equinumerous, i.e. just in case there is a one–one relation which maps all the things falling under one of the concepts of the judgment on to all the things falling under the other concept of the judgment. This criterion has an importance that is independent of whether it is capable of addressing in whole or in part the metaphysical status of numbers as self-subsistent objects, since it states the condition which controls or governs our judgments of equinumerosity in the applications we make of the theory of the natural numbers. Indeed, not only is it a condition of adequacy for any account of our use of the natural numbers for counting that it should imply Frege's criterion of identity, but any account of our conception of the natural numbers which did not extend to an account of their application that is in conformity with the criterion would be regarded as fundamentally flawed.[16]

---

[15] Frege (1884, sects. 43–63).
[16] This is argued in Dummett (1991a, pp. 200–2). It suffices for my purposes that the necessity of a restricted form of the criterion be conceded for the analysis of the applicability of arithmetic. On the character of the restriction, see footnote 5 of Section 9.2, below.

In extra-mathematical applications, the one–oneness of the relation between the objects falling under two concepts will generally depend on empirically confirmable facts about the concepts and the things that fall under them, and the application of the criterion in particular cases will typically decide the truth or falsity of empirical judgments about cardinality. But the question which the application of arithmetic raises regarding its factuality is not whether there are empirical judgments involving the concepts of arithmetic – which of course there are – but whether, when the criterion of identity for number is applied to recognition judgments involving empirical concepts, it is constrained by empirical considerations.[17] In connection with this question, the first thing to notice about Frege's criterion is its generality. Nothing is assumed about the relation that secures the one–one correspondence among concepts whose numbers are the same. The general principle that specifies what is required for sameness of number – namely, the existence of a one–one correspondence – is completely general and is the same whatever the nature of the concepts whose numbers are being compared. The criterion applies to empirical and nonempirical concepts, and it neither invokes nor requires the truth of any empirical hypotheses concerning the character of the relation which comprises the correspondence. The nonfactuality of applied arithmetic follows from the observation that the criterion of identity is not constrained by empirical assumptions of any kind; notice, therefore, that the nonfactuality of arithmetic is not dependent on its reducibility to logic. By contrast, congruence and simultaneity are empirically constrained – by the principles which govern the transport of practically rigid bodies in the case of congruence and by the transmission of light signals in the case of simultaneity. This is the crucial methodological difference between arithmetic and geometry that emerges from Einstein's analysis of the application of geometry and Frege's analysis of the application of arithmetic.[18]

---

[17] Although Carnap's discussion of Frege in *LSL* (sect. 84) acknowledges the importance of the role of numerical concepts in empirical judgments for any account of arithmetic, it misses the point made in the text which concerns the role of the criterion of identity. This is also true of the discussion in Carnap (1939), reviewed earlier (at the conclusion of Section 2.2).

[18] Hume also had the idea that the epistemic status of geometry and arithmetic is illuminated by their respective criteria of identity. But for reasons given in the introduction to this volume, Hume's development of the idea is altogether different from the present proposal. See DePierris (2012) for an instructive discussion of Hume's theory of space, together with a critical review of the recent secondary literature on this topic.

Our discussion of Frege on the application of the Peano – Dedekind theory of natural numbers and of the bearing of Einstein's analysis on the application of Minkowski geometry has shown how the applications of both mathematical theories can be understood to rest on the provision of a criterion of identity. One criterion addresses when the same number is correctly identified as the number of two different concepts, the other addresses when the time of occurrence of two spatially separated events is the same. From the point of view of methodology, the most important difference between geometry and arithmetic is how empirical assumptions enter into the geometrical criterion of identity but not the arithmetical one. In the case of arithmetic, a relation of one–one correspondence determines the condition under which a recognition judgment involving sameness of number holds. The analogue of this in the case of judgments of sameness of time of occurrence is the holding of simultaneity, understood in terms of a relation of signal connectibility. The relation that realizes the correspondence in the arithmetical case is not subject to material conditions or empirical constraints. But since signaling rests on the process of light transmission and light's material properties, it draws on a significant body of theory and experiment. Nothing of this sort occurs in the application of arithmetic. Unlike geometry, arithmetic retains its status as a theory without empirical constraints throughout its application. By contrast, when geometry is applied it is transformed into a component of our theory of the physical world.

A logicist would emphasize that the criterion of identity for number depends on the logical notion of a one–one correspondence. But for a defender of Carnap's thesis the important point is not that the notion is a *logical* one, but that it is not *empirically constrained*. Neither the logical character of the notion of one–one correspondence nor the fact that the notion is not empirically constrained depends on the assumption that the principles of arithmetic in their full generality are reducible to logic.

The parallel between Einstein's analysis of time and Frege's analysis of number can be made even more striking. In classical mechanics, one begins with *the* temporal continuum and derives an analysis of simultaneity judgments from *it*. Although it is true for both classical mechanics and special relativity that the time of occurrence of one event is the same as that of another if and only if the events are simultaneous, in classical mechanics simultaneity is analyzed in terms of *time of occurrence*, the temporal continuum being something that is simply taken for granted in the classical framework.[19] In this respect the situation is reminiscent of pre-Fregean views of number, according to which one starts with the ordinal notion of number and determines that the things falling under one concept are equinumerous with those falling under another just in case their ordinal enumerations have the same final term. Both Frege and Einstein *reverse* the traditional direction of analysis by taking the equivalence relations as primary; they then go on to explain the relevant recognition judgments in terms of the holding of the appropriate relation. But in addition to their reversal of the traditional directions of analysis, Frege and Einstein also provide analyses of these relations, with equinumerosity being explained in terms of one–one correspondence and simultaneity in terms of signal connectibility. What emerges from each analysis is a clarification of the basis for the relevant recognition judgments: in the numerical case they are shown to rest on just those logical notions that are required for the definition of one–one correspondence, and in the temporal case they are shown to rest on experimental and nomological facts regarding signal connectibility.

2.6 CENTRALITY

Let me conclude by returning to an issue that arose in our discussion of the Tractarian strategy by considering the question: 'What role does centrality play in the defense of Carnap's thesis?' Operating under the assumption that analyticity is the *only* basis for the contention that logic and mathematics are nonfactual, it has seemed plausible that Carnap's thesis depends on an underlying confirmationist theory of meaning. Such a theory is compatible with the empiricist proclivities of many proponents of the thesis, and it is clearly possible to argue on its basis that logic and mathematics are non-factual because they are equally confirmed or infirmed by all possible observational evidence. On the surface, this approach appears to facilitate an implementation of the original Tractarian strategy and the idea of basing the explication of factuality on tautologousness: a proposition is *genuine* or *factual* if the observation statements we take to be confirmatory and

---

[19] Consider the *principle of trichotomy*: for any real numbers a and b either $a < b$ or $b < a$ or $a = b$. If one overlooks the possibility that temporal order might depend on an implicit assumption about simultaneity, it can seem that time inherits its order properties from the real numbers and that trichotomy is as much a conceptual truth for moments of time as it is for real numbers. Einstein's analysis made it clear that temporal order cannot be "read off" the real numbers in this way. Thanks to Robert DiSalle for this observation.

infirmatory of it form part of a partition of the class of all possible observation statements. Since a proposition of logic or mathematics is typically held true come what may, it fails to effect such a partition among observation statements, and, on this Quinean development of the Tractarian strategy, is uninformative and without factual content.

A development of the Tractarian strategy in terms of centrality is in some ways congenial to Carnap's own understanding of the basis for his thesis, especially given his emphasis on observation statements for the explication of factual content. But since *any* proposition can be held true come what may, this is at best a defense of the *letter* of the thesis. It does not ensure that observational uninformativeness is the prerogative of the propositions of logic and mathematics, nor does it ensure that they will always be accorded this status. It would therefore be highly problematic if this were the only strategy that could be marshaled in defense of Carnap's thesis: we would be left to distinguish the epistemological status of theories with only the coarse grid of degrees of centrality. As a result we would be unable to recognize a basic difference among mathematical theories that are applied in science, a difference that does not turn on centrality. This is the difference between those mathematical theories whose criteria of identity show them to be empirically constrained when they are applied and, in the sense that is relevant to Carnap's thesis, are therefore such that they have a factual content, and those whose application is not empirically constrained and are therefore without factual content. The distinction between these two types of applied mathematical theory is utterly transparent on the Einsteinian strategy where it is accommodated by considerations that are entirely independent of the notion of centrality.[20]

If the preceding defense of Carnap's thesis is cogent, it shows that there is an epistemologically motivated consideration in favor of separating some applied mathematical theories from the rest of our theoretical commitments. We have seen that this separation can be accomplished without Quine's notion of centrality, and we have seen that it can be accomplished without endorsing the epistemic primacy of the observation language. This vindicates something *close* to Carnap's claim in his long-standing

---

[20] It is an insufficiently appreciated fact that in *LSL* (sect. 82) Carnap anticipated the concluding argument of "Two dogmas" in defense of the holism of confirmation, the possibility of rejecting protocols, and the revision of logical principles in light of new evidence. He did not, however, regard this as grounds for the rejection of his thesis. In terms of the discussion in the text, Carnap perceived that one could acknowledge a role for centrality without necessarily undermining the possibility of a principled distinction between factual and nonfactual components of our conceptual framework. See also Friedman (2006, pp. 47–8) where this point is elaborated.

controversy with Quine that the traditional analytic–synthetic distinction is indispensable to a correct methodological analysis of science.[21] What it vindicates is the traditional distinction's insistence that the difference in epistemological status that should be accorded various applied mathematical theories consists in the factuality of some of these theories, and the nonfactuality of others. It departs from the traditional justification of this distinction by not making it rest on notions of analyticity, convention, or stipulation, but by deriving it from a consideration of the theories' respective criteria of identity.

ACKNOWLEDGMENTS

This chapter has benefited from its presentation to audiences at Concordia University, the University of California/Irvine, and Indiana University. I am indebted to Greg Lavers, Kai Wehmeier, and Amit Hagar for their perceptive comments on one or another earlier version of this chapter and for arranging its presentation at their respective institutions.

---

[21] It also supports Friedman's insistence (2001) that the role of mathematics in physics is seriously distorted when it is regarded as just one more element in the "web of belief."

CHAPTER 3

On extending "Empiricism, semantics and ontology" to the realism–instrumentalism controversy

> If, in some cataclysm, all of scientific knowledge were to be destroyed, and only one sentence passed on to the next generations of creatures, what statement would contain the most information in the fewest words? I believe it is the *atomic hypothesis* . . . that *all things are made of atoms – little particles that move around in perpetual motion, attracting each other when they are a little distance apart, but repelling upon being squeezed into one another.*
>
> (Feynman 1963, pp. 1–2)

The concept of a linguistic framework and the distinction between internal and external questions are the central ideas of Rudolf Carnap's "Empiricism, semantics and ontology."[1] It is not uncommon to encounter the suggestion that reflection on the theoretical and experimental investigations which led to the acceptance of the atomic hypothesis undermines Carnap's distinction between these two types of question and the utility of his notion of a linguistic framework. I believe this is a mistake. There is a natural development of the distinction and the notion of framework choice with which it is paired that is perfectly capable of accommodating this case. I show this by bringing out a subtlety that arises in the extension of the conceptual apparatus of *ESO* to the realism–instrumentalism controversy. When this subtlety is taken into account, the question contested by the nineteenth- and early twentieth-century opponents and proponents of the atomic hypothesis, and successfully addressed by Einstein and Perrin, is readily seen to be internal. Moreover, this formulation of the distinction and the controversy are both independent of Carnap's views on cognitive significance and factual content. I conclude with a presentation and discussion of two formulations of the realism–instrumentalism controversy that are based on Carnap's explication of the factual content of a theory in terms of the notion of its Ramsey sentence.

First published in *Journal of Philosophy* 108 (2011): 647–69.   [1] Carnap (1950); hereafter, *ESO*.

3.1 THE NATURE OF THE PROBLEM

Carnap's primary objective in *ESO* was to clarify the basis for employing a language with expressions for abstract entities. He sought to show that an empiricist should regard the use of such a language as entirely uncontroversial, since it is neither favorable to Platonism nor unfavorable to nominalism when these doctrines are understood in their classical senses as doctrines about the reality or otherwise of abstract entities. The source of the difficulty with the classical doctrines is that they present themselves as theories that inform us about reality rather than as proposals involving the utility of a language form. In so doing they misconstrue the "external" questions they address as comparable in significance to "internal" existence questions. Carnap singles out for special criticism a position he attributes to Paul Bernays,[2] according to which the use of real-number variables for the representation of spatio-temporal coordinates is enough to make a committed Platonist of someone who uses the language of physics as a system of communication "even if he is a strict empiricist who rejects Platonic metaphysics" (*ESO*, p. 215).

ESO thus seeks to clarify the controversy surrounding the legitimacy of a language which employs expressions for abstract entities with this distinction between two types of question. In broad outline, Carnap's basic idea was that internal questions raise genuine existence questions which require a "theoretical investigation." By contrast, external questions raise issues that are addressed by the practical decision to accept a linguistic framework.[3] A key point of Carnap's distinction is that it is characteristic of the answers to genuine internal questions that they are capable of settling "nontrivial"

---

[2] Bernays (1935).

[3] Not all cases involving framework choice are plausibly regarded as purely matters of practical decision. Something more than practical necessity may be involved when a mathematical theory is selected for the formulation of a physical theory. For example, Carnap tells us that:

the external questions of the reality of physical space and physical time are pseudo-questions. A question like 'Are there (really) space-time points?' is ambiguous. It may be meant as an internal question; then the affirmative answer is, of course, analytic and trivial. Or it may be meant in the external sense: 'Shall we introduce such and such forms into our language?'; in this case it is not a theoretical but a practical question, a matter of decision rather than assertion, and hence the proposed formulation would be misleading. Or finally, it may be meant in the following sense: 'Are our experiences such that the use of the linguistic forms in question will be expedient and fruitful?' This is a theoretical question of a factual, empirical nature. But it concerns a matter of degree; therefore a formulation in the form 'real or not?' would be inadequate. (*ESO*, p. 213)

Although of great importance, cases of this last sort raise issues related to the application of the concept of framework choice and the distinction between internal and external questions that fall outside the scope of this chapter. For a preliminary discussion of such cases see Friedman (2001).

questions of existence, such as questions pertaining to the existence of elementary particles with various properties in the case of physics, and the existence of numbers and functions satisfying certain conditions in the case of mathematics. This notion of nontriviality is an informal one: it does not exclude the possibility that a nontrivial answer to an existence question is sometimes an analytic truth. In this case, the answer's nontriviality consists in its having a psychologically informative logical derivation from analytic truths. But when an external question is mistakenly posed as if it were an internal existence question, the answer is characteristically trivial and analytic, even when it purports to address a question of fact rather than a question of logic or mathematics. Notice, therefore, that Carnap's distinction between internal and external questions cannot be straightforwardly identified with the analytic–synthetic distinction.[4]

Carnap gives as a mark of externality the difference between questions that involve the existence of particular elements (or particular classes of elements in the domain of entities over which our individual variables range) from questions about the existence of the entire system of individuals, or some other logical category of entity. He goes on to suggest that any question about a whole system of entities is likely to signal an external question, while a question about the existence of a particular element of the domain is typically an internal one. However, this is at best a rough guide, and one that can be undermined if the domain is assumed to be many-sorted and the sorts are classified in a way that is unfavorable to the intuition Carnap is trying to capture.[5]

Our methods for resolving internal questions follow an intricate and evolving system of rules of proof and evidence, and the answers to these questions consist in the theories and existence claims we come to accept on the basis of such methods. It is a mark – but only a mark – of external questions that these methods, which I will refer to as our "ordinary methods" of proof and evidence, are incapable of resolving them. Because external questions are resistant to our ordinary methods, Carnap holds that they must be answered by a practical decision. But intractability to ordinary methods is insufficient to distinguish external questions from questions which are "merely intractable." For example, there are open questions in number theory that may turn out to be intractable to ordinary

---

[4] This is emphasized by Carnap when he remarks that once "new forms are introduced into the language, it is possible to formulate with their help internal questions and possible answers to them. A question of this kind may be either empirical or logical; accordingly a true answer is either factually true or analytic" (*ESO*, p. 214).

[5] See, for example, Quine (1951b).

methods, but they are unlikely candidates for external questions if our responses to them have little in common with the selection of a linguistic framework. These considerations suggest that the internal–external distinction and the notion of a linguistic framework are interdependent concepts, so that our understanding of them must be achieved in tandem rather than sequentially.

Although intractability to ordinary methods and the property of pertaining to the existence of a domain of quantification are only marks of an external question, they may nevertheless be instructive when considered in conjunction with particular cases. Such a development of the internal–external distinction will be successful to the extent that the classification it determines is capable of providing an illuminating reconstruction of the product of our collective theoretical investigations. However, before considering an example that is particularly instructive in connection with Carnap's distinction, it will be worthwhile to briefly describe a change of linguistic framework that does *not* rest on the distinction between these two types of question. This case of framework change figures prominently in *Logical Syntax of Language*,[6] a work that precedes *ESO* and the formulation of the internal–external distinction. The change of framework it involves is not associated with one or another metaphysical view; nor does it involve the transformation of a philosophical problem into one of choice of language.

By contrast with common examples of what might prove to be intractable problems of number theory, Gödel's discovery of sentences undecidable in first-order arithmetic raises a general question about the characterization of arithmetical truth, one that in *LSL* Carnap proposed to address by a change of linguistic framework. The evident candidate for such a characterization is a definition in terms of deduction from a recursive set of arithmetical sentences in accordance with finitary rules of inference. But Gödel's first incompleteness theorem shows that there is no finitary characterization of the class of arithmetical truths: every such characterization must leave some sentences neither arithmetically true nor arithmetically false, and thus, in the terminology of *LSL*, neither analytic nor analytically false or contradictory. Carnap argued that for anyone who holds that such sentences are not synthetic, Gödel's discovery shows that the definition of analyticity must be emended. This is a purely technical problem, demanding a purely technical solution. Carnap's solution is based on a fundamental theorem of *LSL*, namely 14.3 (*LSL*, p. 40). This theorem

---

[6] Carnap (1934); hereafter, *LSL*.

justifies the adequacy of Carnap's proposed new definition of 'analytic' for the arithmetical fragment of *LSL*'s Language I. (Henceforth, I will take it to be understood that my remarks about *LSL*'s Language I are restricted to its arithmetical fragment. This fragment is a form of primitive recursive arithmetic.) The theorem's content and Carnap's definition may be expressed simply (if somewhat anachronistically) given a small number of preliminary notions.

Suppose we are given a first-order language *L* with equality, where *L* is the arithmetical language customarily adopted to express the first-order theory of the standard model of arithmetic. In *L* it is possible to form a *numeral name* for each natural number *n* using just the nonlogical symbols for zero and successor. A *numerical instance* of a universally quantified sentence $\forall x \varphi(x)$ is the sentence which results from the substitution of a numeral name for the free variable *x* in $\varphi(x)$. ω is the rule of proof which allows us to pass from all the numerical instances of a universally quantified sentence to the universal sentence itself; *Carnap's theorem* is the proposition:

If T is a theory in L strong enough to decide all atomic formulas, then T + ω is complete with respect to all formulas of L.[7]

Carnap's partial solution to the "crisis" caused by Gödel's discovery of essential incompleteness proceeds by dropping the finitariness constraint on rules of proof and proposing a relation of deduction that includes the infinitary rule of proof ω. He then argues (in effect) that the analytic sentences of Language I are given by the deductive closure of a theory which satisfies the conditions of his theorem, where the deductive closure is taken relative to this expanded and *nonfinitary* system of rules of proof. Since primitive recursive arithmetic is one such theory, every sentence of Language I is analytic or contradictory in the sense of Carnap's definitions. Without some such infinitary notion, arithmetical truth would be left indeterminate. *LSL*'s introduction of the ω-rule remedies this indeterminateness, and from the perspective of both *LSL* and *ESO*, its introduction constitutes a change of linguistic framework, whose utility consists in its provision of a determinate concept of arithmetical truth. Indeed, Carnap's proposal is essentially equivalent to the definition of arithmetical truth in terms of truth in the standard model of arithmetic. And because of the extensional equivalence between Carnap's *analytic in the arithmetical fragment of Language I* and *arithmetically true* in the sense of Tarski, Carnap's

---

[7] This exposition of Carnap's theorem was suggested by Isaacson (1992). For the history of ω (the "ω-rule") see Feferman (1986).

definition of analyticity for Language I is often compared with the application of Tarski's definition of truth to the case of arithmetical truth.

As we will see, the idea Carnap pursues in *ESO* is to transform classical philosophical controversies into choices among proposals to adopt one or another language form. This is *not* the point of *LSL*'s discussion of the bearing of Gödel's theorem on the nature of our knowledge of arithmetic. A necessary condition for a solution to this latter question of the sort that Carnap would find acceptable is that we should be in possession of a notion of *analytic in Language I* which partitions the class of arithmetical sentences into two classes: those that are analytic and those that are analytically false. Carnap's definition in terms of ω does this, and this shows that an important *condition of adequacy* for an account of the kind he favors can be met.

LSL's proposed solution to the characterization of arithmetical truth involves what in the terminology of *ESO* would count as a change of linguistic framework. However, it is important to recognize that the transition to a framework that admits a nonfinitary rule of proof neither *solves* nor *dissolves* the philosophical problem of explaining the a priori character of arithmetical knowledge, but serves only to show that Carnap's preferred solution, according to which mathematical knowledge consists in a priori knowledge of analytic truths, has not been automatically ruled out by Gödel's discovery of essential incompleteness. It should also be noted, if only in passing, that it would be a mistake to represent Carnap as following a method of "truth by stipulation," since he does not simply stipulate the conclusion he requires, but gives an informative analysis of arithmetical truth in terms of the ω-rule and then proceeds to demonstrate its adequacy for his purposes. That the theorem which establishes this is not mathematically difficult is quite irrelevant to its interest or to the correctness of Carnap's deployment of it.

There are important points of difference between the notion of framework change that can be extracted from *LSL* and the use to which this notion later came to be put in *ESO* when it was combined with the distinction between internal and external questions. Although both works are concerned with changes of framework, Carnap's goal in *ESO* is to show how the notion of a linguistic framework can be used to *transform* a traditional metaphysical problem into a problem of an altogether different character. Carnap's proposal is that opposed metaphysical positions should not be regarded as differing views about reality, but as expressions of different preferences for the choice of a linguistic framework. According to *ESO*, this is true of the dispute between realists and phenomenalists and, as we noted earlier, of the dispute between Platonists (to whom Carnap sometimes refers as realists in the medieval sense) and nominalists. Questions about reality are susceptible to resolution by ordinary methods. Questions that are advanced as questions about reality, but that are not amenable to such methods, are without "cognitive significance" and should be transformed into questions about the choice of a language form. By contrast, *LSL*'s extension of the linguistic framework to facilitate the definition of analyticity does not transform any philosophical question into one of framework choice. And although it bears on the epistemological question of the nature of arithmetical knowledge, the change in linguistic framework which the extension effects leaves that question pretty much open.[8]

ESO does not list the realism–instrumentalism controversy as an example of an external question, but it would be surprising if, at the time of this paper, Carnap did not regard the question the controversy raises as an external one. When in the book based on his seminar on the philosophy of science Carnap touches on the realism–instrumentalism debate, he suggests an approach to the controversy that is reminiscent of what in *ESO* he argued in connection with the debate between Platonists and nominalists:

the question should not be discussed in the form: 'Are theoretical entities real?' but rather in the form: 'Shall we prefer a language of physics (and of science in general) that contains theoretical terms, or a language without such terms?' From this point of view the question becomes one of preference and practical decision.

A footnote to this passage directs the reader to *ESO* where, we are told,

[the view that] greater clarity often results if discussions of whether certain entities are real are replaced by discussions of preference of language forms . . . is defended in detail.[9]

But the book has little further to say on the subject.

In the balance of this chapter, I will give an account of the application of the distinction between internal and external questions to the realism–instrumentalism controversy that, although in many ways faithful to *ESO*, is not the simple extension that Carnap proposes. Nor will my proposal appeal to the notion of cognitive significance or to a tendentious characterization of metaphysics; to the extent that a development free of these ideas proves possible, Carnap's views about linguistic frameworks and

---

[8] A pre-echo of *ESO*'s approach to metaphysical questions occurs in *LSL*'s discussion of "pseudo-object" sentences. For a recent discussion of *LSL* on pseudo-object sentences as "quasi-syntactic" and its relation to *ESO*'s subsequent conception of metaphysical questions as external, see Carus (2007, pp. 256–61).

[9] Carnap (1995, p. 256); the footnote citing *ESO* and the text quoted above which it accompanies appear only in the second edition of Carnap's book.

the distinction between his two types of question will have been shown to be independent of notions that are generally regarded as problematic.

I will begin with a reconstructive proposal for understanding the application of the distinction between internal and external questions to the nineteenth- and early twentieth-century debate over the atomic hypothesis and the question of the reality of atoms. The discussion of the example of the atomic hypothesis will guide the formulation of my proposal for extending the conceptual apparatus of *ESO* to the realism–instrumentalism controversy. I will then turn to Penelope Maddy's important and influential recent criticism of Carnap on the status of the atomic hypothesis in order to show how, on the extension I propose, the issues Maddy raises can be successfully addressed. I will conclude by explaining how Carnap's Ramsey sentence reconstruction of the language of science bears on the nature of the realism–instrumentalism controversy. We will see that there is more than one way to adapt this reconstruction to the task of extending the internal–external distinction to the controversy between realists and instrumentalists.

3.2 THE REALISM–INSTRUMENTALISM CONTROVERSY AND THE ATOMIC HYPOTHESIS

Even though *ESO* does not discuss the example of the atomic hypothesis and the question of the reality of atoms, one reason why the hypothesis is of special interest is not hard to see: if Carnap's proposals regarding framework choice and abstract entities could be shown to apply mutatis mutandis to a case which, like the case of the reality of atoms, is very implausibly represented as a choice of linguistic framework, then this would be a reason to reject his analysis even for the case of abstract entities, the case for which it was originally intended. The link between the atomic hypothesis and framework choice we will need to explore in order to address this possibility is the realism–instrumentalism controversy and its bearing on the historical debate surrounding the question of atomic reality.

It is natural to view the controversy that separates realists and instrumentalists as one that concerns opposing views of theories which postulate unobservable entities. From this perspective, the question posed by the atomic hypothesis is just a special case of a more general question about the existence of unobservables. It would then seem to follow that if realists and instrumentalists are divided on an external question, proponents and opponents of the atomic hypothesis must also be divided on an external question. But then, according to Carnap's understanding of external questions, the atomic hypothesis raises a question that is covertly about the practical utility of a language form. Hence, to accept the atomic hypothesis and the reality of atoms is to make the practical decision to adopt a particular linguistic framework. But reducing the question of the existence of atoms to a question that can be settled by the adoption of a form of language certainly has the ring of a reductio of Carnap's distinction. Is there an extension of the conceptual apparatus of *ESO* to the realism–instrumentalism controversy and the atomic hypothesis that avoids this consequence?

Suppose we try to develop a more abstract approach to the question which divides realists and instrumentalists and view it as raising a meta-theoretical question about the nature of theories in general. The notion that the issue is a meta-theoretical one is arguably implicit in the assertion that the question driving the controversy is whether theories express truths or are merely instruments that enable us to negotiate among phenomena. The proposal that the atomic hypothesis is just an instrument for prediction would then seem to reduce to the claim that atoms are useful fictions. But it is far from obvious that the question whether atoms are fictions is not more naturally addressed by refining ordinary methods so that they may be brought to bear more decisively on the question of the reality of atoms. In this case the question the atomic hypothesis poses might be better classified as internal rather than external. But then our meta-theoretical interpretation of the realism–instrumentalism controversy does not clearly succeed in representing the question of realism as an external one.

There thus appear to be two natural interpretations of the realism–instrumentalism controversy, neither of which fits comfortably with the ideas of *ESO*. On the interpretation which takes it to raise a question about unobservable entities, the question of the reality of atoms is transformed into a linguistic question. And on the interpretation which takes it to raise a question about the "purely instrumental" character of theories, the question which separates realists and instrumentalists might well be amenable to a theoretical investigation using ordinary methods, and should be classified as internal. A major goal of *ESO* is to show that what often appears as a single question may split into two or more radically different questions. It is therefore consistent with the position set forth in *ESO* to view the existence of atoms as indicative of a general and possibly external question about the nature of theories *and* as an internal question regarding the probable outcome of the application of our ordinary methods of theoretical investigation. Is there a way of developing the internal–external distinction that does not reduce the atomic hypothesis to a linguistic proposal and at the same time preserves the external character of the question of realism versus instrumentalism?

What made the atomic hypothesis importantly different from other theories – and what makes the nineteenth-century controversy surrounding it especially significant for the extension of *ESO* – is the fact that it was widely perceived that the hypothesis might prove intractable to our ordinary methods of theoretical and experimental analysis, with the consequence that there might never be a sufficient basis for deciding it one way rather than another. Given the restrictions to which ordinary methods appeared at one time to be subject, it would have been neither absurd nor unreasonable to doubt whether the question of the existence of atoms could ever be satisfactorily resolved. And indeed, many participants in the nineteenth-century debate over the question of atomism voiced just such a view. The issues were complex and far subtler than whether or not the confirmation of the atomic hypothesis was held to too high a standard by its "verificationist" opponents. The difficulties urged against the hypothesis were sensitive to particular formulations of it, and in the most interesting cases centered on the arbitrariness with which early formulations of the hypothesis allowed key theoretical parameters to be set. Able atomists had able energeticist critics. This left the status of the hypothesis in doubt: in the terminology of the nineteenth-century debate, it left in doubt whether atoms are real or are merely useful fictions. Operating under the not unreasonable but, as it happens, false supposition that the question is intractable, one might well have concluded that the hypothesis is a "metaphysical" one, hence one that poses an external question.[10]

Einstein and Perrin provided the definitive account of how to use the phenomenon of Brownian motion to refine our ordinary methods to a point where they could reasonably be held to establish the reality of atoms. In so doing they vindicated the insights of Léon Gouy and others regarding

---

[10] In this connection it is interesting to recall Poincaré's essay "Hypotheses in physics" where a distinction is drawn between hypotheses which are "natural and necessary," such as the suppositions "that the influence of very distant bodies is quite negligible, that small movements obey a linear law, and that [an] effect is a continuous function of its cause," and those that should be qualified as "neutral." Poincaré argued that the atomic hypothesis falls into the second category: "In most questions the analyst assumes, at the beginning of his calculations, either that matter is continuous, or on the contrary, that it is formed of atoms. He might have made the opposite assumption without changing his results. He would only have had more trouble to obtain them – that is all." Regarding an analyst whose calculations assume a neutral hypothesis like the atomic hypothesis, Poincaré continues, "If . . . experiment confirms [such an analyst's] conclusions, will he suppose that he has proved, for example, the real existence of atoms?" The clear suggestion of "Hypotheses in physics" is that no respectable analyst would draw such a conclusion because he would recognize that the hypothesis is merely a convenient aid to calculation. (All quotations are from Poincaré 1900, p. 135.) The idea that the atomic hypothesis is at best a convenient aid to calculation was by no means Poincaré's final view on the subject. See especially Poincaré (1912a) and (1912b), both of which are discussed in Chapter 4, below.

the relevance of Brownian motion and the kinetic theory of gases to the verification of a physical–chemical hypothesis about the constitution of matter.[11] The Einstein–Perrin analysis achieved two broad successes: (i) It led to new determinations of Avogadro's number, the molecular diameter, and numerous other parameters by exploiting independent sources of evidence and new experimental techniques. But in addition, (ii) Perrin's confirmation of Einstein's quantitative analysis showed that Brownian motion refutes the strict (nonstatistical) validity of Carnot's principle, and hence that of the Second Law of Thermodynamics. The strict validity of Carnot's principle was not only a central pillar of the energeticist argument in favor of its alternative to atomism, it was a central pillar of its critical assessment of atomism.[12] As Michael Gardner[13] has observed, it is the combination of these two successes that explains why the Einstein–Perrin analysis of Brownian motion was a development of previous work with such decisive significance for resolving the question of atomic reality. This is therefore a case where what appeared to some to be a question that could only be resolved by a practical decision was shown to be entirely amenable to an innovative extension of ordinary methods. In short, Einstein and Perrin undermined whatever justification there may have been for regarding the question of the atomic hypothesis as external because intractable.

It is clearly desirable that an extension of *ESO* which represents the realism–instrumentalism controversy as one involving an external question should be compatible with the idea that the Einstein–Perrin analysis demonstrated the existence of atoms. What such an extension might never-theless call into question is whether the successful deployment of ordinary

---

[11] The historical development is described in detail in Nye (1972). Nye remarks that Gouy was the only major French physicist to have dealt with the problem of Brownian motion before Perrin. Gouy was prescient but he did not have the benefit of the quantitative analysis of Einstein and others.

[12] Maxwell's Demon hypothesis made it clear that the molecular hypothesis must allow for the possibility that the Second Law holds only statistically. What was striking about Perrin's confirmation of Einstein's quantitative analysis of molecular motion was its demonstration, via the phenomenon of Brownian motion, of the violation of Carnot's principle at the molecular dimensional scale. The topic is discussed by Perrin:

> [Carnot's] principle asserts ... that in a medium in thermal equilibrium no contrivance can exist capable of transforming the calorific energy of the medium into work ... [t]hat is to say, [into] doing work without taking anything in exchange and without external compensation. [Such] *perpetual motion of the second kind* is held to be impossible. Now we have only to follow, in water in thermal equilibrium, a particle denser than water, to notice that at certain instants it rises spontaneously, thus transforming a part of the heat of the medium into work. If we were no bigger than bacteria, we should be able at such moments to fix the dust particle at the level reached in this way, without going to the trouble of lifting it and to build a house ... without having to pay for the raising of the materials. (Perrin 1916, pp. 86–7)

[13] Gardner (1979).

methods in support of the hypothesis can be parlayed into a general argument in favor of realism over instrumentalism. The intuition that it cannot is based on the idea that when realism and instrumentalism are understood as meta-theoretical claims about the interpretation of theories, it is generally assumed that both realists and instrumentalists agree on what theories the evidence favors and on the decisiveness with which it favors one theory over another. Understood as an external question about realist and instrumentalist interpretations of theories, the question of the existence of atoms is not one ordinary methods address. But granting this does not preclude the possibility that there is a perfectly good *internal* question about the reality of atoms that *is* addressed by such methods.

I have placed special emphasis on the idea that the *point* of Carnap's distinction is to clarify situations in which what appears to be a single question about a hypothesis splits into two or more very different questions, each having a methodology appropriate to its resolution. In keeping with our discussion of the atomic hypothesis, it is natural to distinguish those questions that are amenable to ordinary methods of theoretical investigation from those that are not. This brings us to the subtlety that arises in the application of the internal–external distinction to the realism–instrumentalism question that I noted earlier: among the questions that are not amenable to ordinary methods, we should distinguish between questions for which ordinary methods fail to *stabilize* on a particular hypothesis – as at one time seemed would be the case for the atomic hypothesis – and questions that *fall altogether outside the purview* of such methods. Both are possible cases of external questions. But it is only the case where a question is judged to be external because it is believed that ordinary methods will never stabilize on a verdict that involves a speculative guess about the future of science. The view that atoms should be regarded as useful fictions was based upon just such a speculative guess. As we saw, such guesses can be mistaken, with the consequence that the views they are taken to support – such as the view that atoms are merely useful fictions – become groundless.

The situation is different with questions which fall outside the scope of what ordinary methods can deliver. That there should be external questions in this sense is clearly essential to the point of Carnap's distinction. But the elaboration and defense of the distinction should be compatible with a variety of approaches to establishing that there are such questions. One way to establish this is to appeal to the idea that the metaphysical question which separates realists and instrumentalists is one that concerns the language within which our understanding of claims regarding the truth of theories is expressed. On this way of framing the debate, the instrumentalist maintains that truth should be identified with warranted assertability, while the realist denies any such identification. Theoretical and experimental investigations based on ordinary methods, however decisive, are irrelevant to whether the evidence garnered by such methods can compel us one way or the other on which language form to adopt.[14]

Such an account of the external question which separates realism from instrumentalism is reminiscent of the view of realism and its critics that for a long time has been defended by Michael Dummett.[15] Rather surprisingly, it represents a possible extension of the apparatus of *ESO* to the realism–instrumentalism controversy that is faithful to many of Carnap's basic ideas. However this "Dummettian" extension does not rest on a notion of cognitive significance; nor does it rest on a pejorative characterization of metaphysical questions as meaningless or as "pseudo-questions." Nevertheless, Dummett's proposal that the decision about which notion of truth to adopt should turn on our choice of logic is certainly consonant with Carnap's view that our choice of logic and our answers to external questions are alike matters of practical decision concerning framework choice. Where Carnap and Dummett differ is over Dummett's suggestion that both logic and metaphysics can be derived from an underlying theory of meaning, and the correlative idea that this observation is capable of mitigating the conventionality that for Carnap attaches to answers to external questions and our choice of logic.

We are therefore in a position to conclude that it is at least consistent with the framework of *ESO* that (i) the reality of atoms raises an internal question, but (ii) this recognition is compatible with an understanding of the realism–instrumentalism controversy according to which the question separating realists and instrumentalists cannot be decided by ordinary methods. At the same time (iii) it can happen that the internal or external nature of an existence question, such as the one raised by the atomic

---

[14] Carnap has a remark on the nominalism–realism controversy that comes very close to this formulation of what constitutes a question as external:

> I cannot think of any possible evidence that would be regarded as relevant by both philosophers, and therefore, if actually found, would decide the controversy or at least make one of the opposite theses more probable than the other . . . Therefore I feel compelled to regard the external question as a pseudo-question. (*ESO*, p. 219)

> On my formulation it is the clarification of the *externality* of the question that the decisiveness of ordinary methods is capable of illuminating; I should also emphasize that on my formulation, there are different ways in which ordinary methods can fail to decide a thesis. (Thanks to Thomas Uebel for bringing this passage to my attention.)

[15] See, for example, Dummett (1991b).

hypothesis, depends on the capacity of ordinary methods to stabilize on a verdict for the hypothesis in question.

3.3 A RECENT CRITICISM OF *ESO*

In *Second Philosophy: A Naturalistic Method*,[16] Penelope Maddy remarks on the wide range of things that according to *ESO* are candidates for linguistic frameworks; not only is there a linguistic framework for abstract entities, but there is also one for atoms:

> [T]here is a linguistic framework for talking about observable things and events in space and time, which Carnap calls 'the thing language' . . . There are . . . linguistic frameworks for numbers, including variations with weaker or stronger logics; linguistic frameworks for unobservable entities like atoms, with evidential rules spelling out what observations would count as evidence for and against; mathematics-heavy linguistic frameworks for relativity theory, for quantum mechanics, and so on . . . Questions of existence, of logic, of evidence and truth, he insists, can only be asked within a framework, as it is the framework alone that gives them cognitive significance. (Maddy 2008, pp. 68–9)

It is important to recognize that the suggestion that Carnap requires a special linguistic framework for atoms, or an "atom framework," in order to raise the issues about the constitution of matter that the atomic hypothesis addresses is an extrapolation from *ESO*. Indeed, as we have emphasized, *ESO* does not even *mention* the philosophical controversy that is most closely associated with the atomic hypothesis. And when, in *ESO*, Carnap does consider an extension of the thing language, it is not an extension to the atom framework he considers, but an extension to a framework that includes space-time points. This question is in turn immediately reduced to one involving the mathematical resources of the linguistic framework and whether spatio-temporal coordinates should be given by integers or by rational or real numbers.[17]

Maddy's initial objection, which is directed at the claim that the reality of atoms is to be resolved by the conventional adoption of a linguistic

---

[16] Maddy (2008).

[17] The step from the system of things (which does not contain space-time points but only extended objects with spatial and temporal relations between them) to the physical coordinate system is again a matter of decision. Our choice of certain features, although itself not theoretical, is suggested by theoretical knowledge, either logical or factual. For example, the choice of real numbers, rather than rational numbers or integers, as coordinates is not much influenced by the facts of experience but mainly due to considerations of mathematical simplicity. (*ESO*, p. 212)

I am grateful to Michael Friedman for calling my attention to this passage.

framework, is therefore an objection to an extrapolation from Carnap's remarks, rather than an objection to an explicit thesis of *ESO*:

> [L]et's suppose we've adopted a linguistic framework for simple scientific observation and generalization – perhaps an elaboration of the thing language – and we're wondering whether or not to embrace a new range of entities, say atoms. As our current language has no terms for such things, no predicate 'is an atom,' no evidential rules with which to settle questions of their existence or nature, Carnap holds that this is not a question that can be asked or answered internally, that we must step outside our linguistic framework and address it pragmatically, as a conventional decision about whether or not to adopt a new linguistic framework . . .
>
> [Now t]he meticulous and decisive work of Jean Perrin on Brownian motion came as a welcome surprise. In circumstances like these, where the new evidential rules are such elusive and hard-won scientific achievements, [one] is unlikely to agree with Carnap that their adoption is a purely pragmatic matter, a conventional choice of one language over another. [Rather] the development of the Einstein/Perrin evidence was of a piece with [our] standard methods of inquiry . . . [Even if] the empirical study of human language use might justify some notion of purely linguistic truth, [it is doubtful] that a distinction so grounded would put the relevance of Einstein/Perrin's work to the existence of atoms on the linguistic side of the ledger. (Maddy 2008, pp. 71–3)

According to Maddy, the distinction between internal and external questions is so poorly drawn, and the concept of a linguistic framework so unconstrained, that there is no basis for objecting to a classification of the question of the existence of atoms as wholly one of framework choice. But then the notions of external question and linguistic framework yield a clearly incorrect interpretation of the significance of the work of Einstein and Perrin. And since their work constitutes one of the most significant chapters in the history of the atomic theory, this is sufficient to justify rejecting the method of rational reconstruction that is based on these notions.

I hope it is clear that such an objection overlooks the multifaceted role ordinary methods play in the distinction between internal and external questions, and I hope it is also clear that such an objection imputes too simple a picture to the part played by framework choice in settling existence questions. *ESO* is *not* compelled to represent the atomic hypothesis as raising a question of framework choice in Maddy's sense, let alone as *merely* raising a question of framework choice in this sense. And while it is, of course, true that on Carnap's view the question regarding the existence of atoms which was addressed by Einstein and Perrin demands a "linguistic framework" for its formulation and eventual resolution, this is not a thesis anyone should dispute. The sense in which the atomic hypothesis may be held to require a linguistic framework need amount to nothing more than the demand for a language within which questions about the constitution of matter can be meaningfully posed. Such a demand does not imply – nor is it intended to imply – that the question addressed by Einstein and Perrin is about language choice. The conceptual resources of such a framework can be considerably less than the totality of possible theories with which we might speculate about the constitution of matter, and in so far as the framework must allow for a variety of possible answers, there must also be a sense in which it is more than the conceptual resources of any particular theory. As we have seen, it is entirely compatible with the conceptual apparatus of *ESO* for there to be an altogether different explanation of why the question of the reality of atoms may sometimes take the form of an external question. According to our proposal, the practical decision involves a choice between frameworks which agree in admitting the possibility of quantification over atoms but differ in their notions of truth: in one truth is stronger than warranted assertion, while in the other it coincides with warranted assertion.

Now Maddy also considers the possibility that the question of the reality of atoms is a special case of the external question that separates realists from instrumentalists. In a discussion that is explicitly directed at *LSL* and Carnap's "principle of tolerance," but which is also intended to apply to *ESO*'s emphasis on the pragmatic character of the choice of a linguistic framework, Maddy writes:

[T]he "scientific realist," who believes that atoms exist, reconstructs scientific theorizing along the lines of the atom language, once again with evidential rules strong enough to allow us to justify our belief in what we cannot directly observe. For his counterpart, the "instrumentalist," the projected evidential rules are weaker, and we can never know such things. Carnap's point is that what seems a serious philosophical question, the locus of heated debate – can we know ... that atoms exist? – actually hinges on no more than a conventional choice of rational recon-struction.[18] (Maddy 2008, p. 79)

---

[18] This last remark is clarified in a footnote:

To put [it] another way: Carnap holds that the actual language the "realist" and the "[instrumentalist]" are speaking is hopelessly imprecise, amorphous; that it only makes sense to speak of evidence in the context of a rational reconstruction; that their shared language can be reconstructed in various ways, some conducive to realism and some to [instrumentalism]; that there's no fact of the matter at stake in the choice between these possible reconstructions, only a pragmatic choice. (Maddy 2008, p. 79)

Maddy's discussion focuses on realism, instrumentalism, and skepticism; I have made it uniformly about realism and instrumentalism, and have omitted any reference to skepticism. My insertion of '[it]' in the first line of Maddy's footnote corrects an obvious typo.

But there is a fundamental difficulty with this way of subsuming the question of the existence of atoms under the external question that is contested by realists and instrumentalists: by representing Carnap's position as one that makes the existence of atoms relative to the choice of a *realist* or an *instrumentalist* rational reconstruction of the atomic theory and its evidential base, Maddy effectively forecloses the possibility that for Carnap the question of the reality of atoms splits into an internal question that is amenable to ordinary methods and an external question that is not amenable to such methods. Let me conclude this section with a summary of what I take the discussion so far to have established.

Maddy's discussion is directed at showing that both Carnap's distinction and its point are unclear. But on our proposed extension of *ESO* to the realism–instrumentalism controversy, there should be little for Maddy's "Second Philosopher" to find fault with. The clarity of the distinction between internal and external questions and its extension to the realism–instrumentalism controversy rests only on the clarity of the notion of ordinary methods of proof and evidence, and this is a notion whose clarity the Second Philosopher should be willing to concede. As for the *point* of Carnap's distinction, it aims to clarify the scope of our ordinary methods and the character of those questions, if any, which transcend them. But the clarification of the scope of our ordinary methods is a goal that the Second Philosopher also shares; at most she is likely to disagree with Carnap over whether there are questions that transcend our ordinary methods and that therefore demand a practical decision regarding the choice of a language form. The nature of truth is plausibly such a question. But on this under-standing of Carnap's distinction, the question addressed by Einstein and Perrin is no more one of language choice than it is for the Second Philosopher. And in so far as the question of the reality of atoms is not decided by a choice of a linguistic framework, as on Maddy's account of Carnap it must be, her implicit conception of how the apparatus of *ESO* extends to the realism–instrumentalism controversy should be resisted.

3.4 CARNAP AND THE REALISM–INSTRUMENTALISM CONTROVERSY

Is there a place subsequent to *ESO* where Carnap discusses the question of the reality of atoms from the perspective of the controversy between realists and instrumentalists? And if so, how do his views compare with the view of the hypothesis and the controversy presented here?

In his response to Hempel, in his "Replies and systematic expositions,"[19] Carnap considers a question that is in all relevant respects the same as the question of the reality of atoms, and what he says in connection with it bears comparison with our reconstruction of the distinction between internal and external questions and its relevance to the atomic hypothesis. After explaining how the Ramsey sentence of a theory succeeds in eliminating theoretical *terms*, Hempel observed that the Ramsey sentence does not eliminate reference to theoretical *entities*.[20] Hence in the case of at least some theories, any domain over which the Ramsey sentence of the theory is interpreted must be an extension of the domain of observable entities. Is it not therefore the case that a proposed reconstruction in terms of Ramsey sentences favors realism? And is this not in tension with Carnap's suggestion, which we quoted earlier, that the question of realism reduces to the linguistic question of whether the reconstruction of the language of physics should include theoretical terms?[21]

Carnap's reply depends on the assumption that a theory and its Ramsey sentence have the same *factual content*, where this means that a theory's Ramsey sentence can replace the original theory in any argument to a conclusion formulated in the theory's observation vocabulary. Any such argument requires for its formulation only the observation vocabulary of the original theory together with the theory's logical and mathematical apparatus. But the terms belonging to the observation vocabulary are just terms of the thing language whose criteria of application have been refined and more precisely regimented. Carnap concedes that since both realists and instrumentalists accept the thing language – with its vocabulary regimented and its logical and mathematical resources suitably enriched – they must both also accept any extension of the domain of observable entities that is demanded by the Ramsey-sentence reconstruction of the theory. As we

---

[19] Carnap (1963b).

[20] Hempel (1958/1965, sect. 9). The notion of the Ramsey sentence of a finitely axiomatized theory assumes that the theory's nonlogical vocabulary has been partitioned into observation and theoretical terms; the Ramsey sentence is then formed by existentially generalizing away the theoretical terms. The philosophically contentious thesis that Ramsey-sentence reconstructions share with the logical-empiricist conception of theories is the assumption that our understanding of the theoretical vocabulary is importantly incomplete and problematic in a way that our understanding of the observation vocabulary is not. A Ramsey-sentence reconstruction takes this incompleteness of the theoretical vocabulary to justify the claim that the Ramsey sentence captures the theory's "factual" content. See Chapters 4, 5, and 7, below, for further discussion of the philosophical interest of Ramsey sentences and an exposition of their logical properties.

[21] Carnap's Ramsey-sentence reconstruction was presented as early as 1959 in Carnap's address to the American Philosophical Association in Santa Barbara, and Carnap was in correspondence with both Hempel and Herbert Feigl on these matters much earlier than 1959. See the historical notes to the publication of the transcription of Carnap's address in Psillos (2000).

just saw, this is precisely what Hempel suggests is too strong a conclusion to be acceptable to an instrumentalist.

Carnap argues, against Hempel, that the Ramsey sentence does not favor realism on the ground that while it

> does indeed refer to theoretical entities by the use of abstract variables . . . it should be noted that these entities are . . . purely logico-mathematical entities, e.g., natural numbers, classes of such, classes of classes, etc. Nevertheless, [the Ramsey sentence] is obviously a factual sentence. It says that the observable events in the world are such that there are numbers, classes of such, etc., which are correlated with the events in a prescribed way and which have among themselves certain relations; and this assertion is clearly a factual statement about the world. (Carnap 1963b, p. 963)

One possible interpretation of these remarks – which constitute the core of Carnap's reply to Hempel – is that the Ramsey sentence of a theory admits an interpretation, in the sense of a specification of a domain for its bound variables, that should be acceptable to an instrumentalist; at the same time, it does not preclude a realist interpretation. On this understanding of the controversy, realism and instrumentalism should not be viewed as alternative rational reconstructions of the language of science, but as alternative interpretations of its *preferred reconstruction* in terms of the notion of a Ramsey sentence. However different realism and instrumentalism may appear as interpretations of a theory's Ramsey sentence, they concur in their representation of the theory's factual content. And since questions of factual content exhaust the internal questions a theory is capable of raising, realism and instrumentalism can differ at most on an external question.

Such an interpretation of Carnap's reply to Hempel suggests the following picture of theories which, like the atomic hypothesis, involve unobservable entities. Assuming realists and instrumentalists both accept the Ramsey-sentence reconstruction of any such theory, they must agree on the theory's factual content. The Ramsey-sentence reconstruction makes a concession to realism by countenancing extensions of the domain of observable entities; at the same time, it concedes to instrumentalism the alternative of understanding any such extension entirely in terms of the number- and set-theoretical resources of the representational framework within which the theory is formulated. That both interpretations are equally legitimate is justified by the explication of factual content. As for the external question on which they differ, the formulation that naturally suggests itself is that the question is whether, as the realist holds, some theories, like the atomic hypothesis, describe a reality that transcends our observation in a fundamental way, or whether, as the instrumentalist maintains, such theories merely present us

with an elaborate mathematical apparatus with which to successfully anticipate the phenomena we observe. The decision between these alternatives is a
practical one because both realists and instrumentalists accept the Ramsey
sentence as a correct reconstruction of factual content. As with our Ramsey
sentence-independent account of the external question that separates realism
from instrumentalism, the decision is expressed by a choice of framework,
namely, by a choice of *interpretational* framework for evaluating the truth of
the Ramsey sentence of the theory. Finally, by its elimination of theoretical
vocabulary, the Ramsey sentence confirms the intuition that internal theoretical questions regarding unobservables can be posed in the thing language,
a language which is acceptable to both realists and instrumentalists.

For anyone who accepts Carnap's Ramsey-sentence reconstruction of the
language of science, this should appear a defensible account of the realism–
instrumentalism controversy as an external question. However, it is not
entirely clear that it is the defense Carnap is proposing, and several considerations can be raised against the claim that it accurately represents his view.
First, there is the passage[22] in which Carnap locates the difference between
realism and instrumentalism in the acceptance or rejection of the use of
theoretical vocabulary. By contrast, the proposed account has both realists
and instrumentalists concurring in their acceptance of a reconstruction that
declines the use of theoretical vocabulary. Secondly, and more importantly,
in his response to Hempel Carnap does not suggest that the variables of the
Ramsey sentence *may* be differently interpreted, but declares that their
correct interpretation is in terms of "purely logico-mathematical" entities.[23]
Thirdly, Carnap argues that physicists may, if they so choose,

evade the question about [the existence of electrons] by stating that there are certain
observable events, in bubble chambers and so on, that can be described by certain
mathematical functions, within the framework of a certain theoretical system . . .
[T]o the extent that [the theoretical system] has been confirmed by tests, it is
justifiable to say that there are instances of certain kinds of events that, in the
theory, are called 'electrons.' (Carnap 1966 and 1995, pp. 254–5)

Thus, for Carnap, the question of realism concerns the convenience that may
or may not attach to the use of theoretical vocabulary, much as in *ESO* the
question of Platonism is transformed into one that rests on the convenience
afforded by a language which contains expressions for abstract entities.

These considerations suggest that even if our proposed account of the
realism–instrumentalism controversy in terms of Carnap's Ramsey-sentence

---

[22] Carnap (1995, p. 256).   [23] I am indebted to Anil Gupta for emphasizing this point to me.

reconstruction of theories may be defensible, it is very likely not his account. To be sure, it shares with Carnap's proposal the idea that the question separating realists and instrumentalists is not factual, but it construes the question of framework choice differently: on our application of his Ramsey-sentence reconstruction, realists and instrumentalists evaluate the truth of the Ramsey sentence relative to different interpretational frameworks. For Carnap, realists and instrumentalists are divided over whether to use theoretical vocabulary, with realists endorsing its use and instrumentalists refraining from its use. Carnap's considered view of the controversy is entirely continuous with his approach in *ESO* to the Platonism–nominalism controversy: Carnap seeks to legitimize the use of theoretical vocabulary without incurring a commitment to realist metaphysics, just as, in the case of Platonism, he sought to show that the use of a language with expressions for abstract objects should not be taken to involve a commitment to Platonic metaphysics.

But in light of our discussion of the atomic hypothesis, Carnap's deployment of his Ramsey-sentence reconstruction should strike us as unsatisfactory: it portrays the question of the reality of unobservables as metaphysical; hence, one that should be transformed into a question of preference for theoretical vocabulary. But then it is difficult to see how the question of the reality of atoms – which are just a special case of unobservables – should not also be regarded as a question of linguistic preference. This is to relinquish at the level of the realism–instrumentalism debate everything we struggled to sustain in connection with the significance of the work of Einstein and Perrin, since it leaves Carnap open to the charge that the question the atomic hypothesis raises can be settled by a choice of language.

Our proposal for applying Carnap's Ramsey-sentence reconstruction to the realism–instrumentalism controversy – our Ramsey sentence-*dependent* proposal – begins from the premise that for both realists and instrumentalists the Ramsey sentence of a theory yields its correct reconstruction. If this is granted, then the question which separates realists and instrumentalists must be an external one: the question on which they differ concerns the preferred interpretational framework for evaluating the truth of the Ramsey sentence. This proposal leaves intact the intuition that the decision whether to accept the atomic hypothesis is not a practical one about the desirability of using theoretical vocabulary. The contrast between it and Carnap's deployment of the Ramsey sentence reflects a difference in conception regarding the problems of abstract and theoretical entities. Pre-analytically, there is an underlying presumption in favor of the theoretical entities of physics; this is evident from the epigraph for this chapter. Such a presumption is far less clear in the case of abstract objects and purely mathematical theories. For an adherent of Carnap's

Ramsey-sentence reconstruction, the extension of *ESO* that I am proposing respects this difference. It does so not by arguing against Platonism in mathematics, but by presenting an account of the realism–instrumentalism controversy that, unlike Carnap's, does not elide the difference between the issues realism poses in the case of physics and the questions it raises in the case of mathematics.

By contrast, our Ramsey sentence-*independent* proposal for extending the distinction between internal and external questions to the realism–instrumentalism controversy does not rest on Carnap's explication of factual content, but appeals only to our ordinary methods of proof and evidence and the idea that when our methods stabilize in the evaluation of alternative hypotheses, the issue that separates realists from instrumentalists concerns the nature of truth and its relation to warranted assertion. As we have just seen, even an account of the debate in terms of his explication of the factual content of a theory by its Ramsey sentence is strengthened when it is seen to turn on issues involving truth: in the present case, on different conceptions of how to understand the truth of the Ramsey sentence. I would like to conclude by emphasizing once again that the core ideas of *ESO* – the distinction between internal and external questions and the notion of a linguistic framework – are independent not only of the notion of cognitive significance, but are independent also of the central tenets of Carnap's Ramsey-sentence reconstruction of the language of science. The most fundamental of these tenets concerns the nature of the epistemic primacy which is accorded the observation language. It is not my purpose here to assess this tenet.[24] I will only remark that it is a highly nontrivial assumption, and for anyone attracted to the ideas of *ESO*, it should be a welcome consequence that it is an assumption which their defense does not require.

ACKNOWLEDGMENTS

I am especially indebted to Michael Friedman and Anil Gupta for extended discussions on the topics dealt with here. Thomas Uebel and André Carus made valuable suggestions for which I am very grateful. I wish to thank Melanie Frappier and Jeffrey Bub for sharing their translation of a hitherto untranslated paper of Poincaré's. For their perceptive comments on earlier drafts I would like to thank Penelope Maddy, Greg Lavers, Andrew Lugg, Gerry Callaghan, and Aaron Barth. The Social Sciences and Humanities Research Council of Canada and the Killam Trusts supported my research.

---

[24] It is assessed in Chapters 4 and 7.

CHAPTER 4

Carnap's analysis of realism

Michael Friedman has recently observed[1] that Carnap's reconstruction of theories and view of the realism–instrumentalism controversy are clarified when they are considered in light of an objection to Bas van Fraassen's constructive empiricism, an objection that exploits a variant of Hilary Putnam's well-known model-theoretic argument.[2] After reviewing this objection, I will explain Friedman's observation. I will then turn to a discussion of Carnap's application of his "Ramsey-sentence reconstruction of theories" to the realism–instrumentalism controversy, where I will show how a purely logical distinction between two types of theory is particularly useful for understanding the peculiarities and limitations of Carnap's view.[3] Although I fully endorse Carnap's philosophical neutrality on metaphysical questions of ontology, I believe that the question which separates realists and instrumentalists is neither metaphysical nor a question about which we should be neutral. I will, however, formulate a question that separates realists and anti-realists that is plausibly represented as a metaphysical question – a question about the notion of truth which is applicable to certain empirical theories. This is a question on which Carnap's Ramsey-sentence reconstruction is *not* neutral, but favors anti-realism. I conclude with a critical discussion of this form of anti-realism, and in the appendix to this chapter, I comment on van Fraassen's empiricist structuralism.

---

[1] In Friedman (2012).

[2] This objection was first raised in Demopoulos (2003a, sect. 5), which appears as Chapter 6 of this volume. By Putnam's model-theoretic argument I mean the argument presented in Putnam (1977), Putnam's presidential address to the American Philosophical Association. As noted in Demopoulos and Friedman (1985) – Chapter 5 of the present volume – Putnam's argument shares a formal affinity with the criticism of Russell's causal theory of perception presented in Newman (1928).

[3] The distinction is elaborated in Chapter 7 below.

4.1 CONSTRUCTIVE EMPIRICISM AND ITS DISCONTENTS

In *The Scientific Image*,[4] van Fraassen formulated a theory of theories and a theory of scientific knowledge that has been highly influential in the philosophy of science of the last twenty or so years. Van Fraassen called his theory of scientific knowledge *constructive* empiricism – to distinguish it from empiricist approaches that are defined by some form of reductionism, whether to sense-data or to observation statements – and he called his theory of theories, the *semantic* view of theories, to contrast it with the preoccupation of earlier approaches – especially those of Carnap and the logical empiricists – with the *syntax* of the language of science.

The Scientific Image appeared when the logical-empiricist view of a theory as an abstract calculus, a calculus whose connection to experience is mediated by a "partial interpretation," was being displaced by an emerging realist consensus. One of the main goals of van Fraassen's book was to establish that constructive empiricism represents a radical departure from logical empiricism that is demonstrably preferable to realism: when measured against any of its realist alternatives, constructive empiricism is readily seen to be the more rational position to adopt regarding the existence of theoretical – i.e. unobservable – entities. The criticism of constructive empiricism I developed in Demopoulos (2003a) exploited the fact that the argument van Fraassen advances in defense of this claim requires a distinction between empirically adequate and true theories that is undermined by other aspects of his view. Let me briefly recall the main thread of this objection to constructive empiricism.

Assuming the semantic view of theories, according to which a theory is identified with a class of models, constructive empiricism is characterized by a central tenet and a pair of definitions. The *central tenet* is that it is always more rational to accept a theory which postulates unobservable entities as empirically adequate than it is to believe it to be true. The *definitions* are: (i) that to *accept a theory as empirically adequate* is to hold that the phenomena can be "fitted into" a model belonging to the class of models that *is* the theory; and (ii) that to *believe a theory true* is to hold that it contains (a representation of) the world among its models. More precisely, a theory is *empirically adequate* if the *observable* phenomena can be isomorphically imbedded into one of its models. The notion of truth that is applicable to theories is a simple extrapolation from this definition of what it is for a theory to be empirically adequate to a definition that involves observable

---

[4] Van Fraassen (1980).

and *unobservable* phenomena: a theory is *true* – and not merely *empirically adequate* – if there is a model belonging to the theory which isomorphically represents the situation of both observable and unobservable entities.

Realists believe that some theories which postulate unobservable entities are true. Constructive empiricists do not deny that theories – i.e. the models theories comprise – include representatives of unobservables among the elements of their domains; they are merely agnostic regarding a theory's claims about them. Thus, van Fraassen tells us that constructive empiricism holds

> that physical theories do indeed describe much more than what is observable, but what matters is empirical adequacy, and not the truth or falsity of how they go beyond observable phenomena ... To present a theory is to specify a family of structures, its *models*; and secondly, to specify certain parts of those models (the empirical substructures) as candidates for the direct representation of observable phenomena. The structures which can be described in experimental and measurement reports we can call *appearances*; the theory is empirically adequate if it has some model such that all appearances are isomorphic to empirical substructures of that model. (van Fraassen 1980, p. 64)

The difficulty with this position is that so long as a theory consists of empirically adequate models with domains of the right cardinality, constructive empiricism implies that a theory is true if it is empirically adequate. Hence on constructive empiricism's own account of the truth of theories that go beyond the phenomena by postulating unobservable entities, there is virtually nothing to choose between a true theory and one that is merely empirically adequate. This is contrary to the idea – central to van Fraassen's defense of the claim that constructive empiricism is preferable to realism – that there is a substantive difference between believing a theory true and accepting it as merely empirically adequate.

To establish this observation, we need to show that in van Fraassen's sense of 'theory' any theory that "saves the phenomena" contains among the models it comprises one that corresponds to the world. We can show this if we can assume that the domain of one of the theory's empirically adequate and partially abstract models – a model whose existence follows from the consistency of the theory[5] – is in one–one correspondence with the domain of observable and unobservable entities. Let M be a partially abstract model

---

[5] See van Benthem (1978), where this folklore result occurs as:

> Lemma 3.2: For any L_o-structure M, if M is a model of the L_o-consequences of T, then there exists an $L_o \cup L_t$-structure N such that N is a model of T and the reduction of N to L_o is an elementary extension of M.

that satisfies this cardinality condition. Now *define* the theoretical relations on the domain W of the model corresponding to the world so that, for example, the theoretical relation R^W holds of $<a, b>$ if and only if R^M holds of $<\varphi^{-1}a, \varphi^{-1}b>$, where φ is any one–one correspondence from M on to W that reduces to the identity map on $M \cap W$ – the observable part of M and W. This transforms a theory that merely saves the phenomena into a true one, since it ensures that a model corresponding to the world is a member of the class of models that is the theory. And it undermines the tenet that it is always more rational to accept a theory as empirically adequate than to believe it true. It undermines this tenet because it shows that for constructive empiricism the difference between an empirically adequate theory and a true one – and thus the basis for favoring constructive empiricism over realism – rests entirely on an assumption about cardinality. Since it is evident that the rationality of the choice between constructive empiricism and realism should not depend on the cardinality of W, this shows that the combination of constructive empiricism plus the semantic view of theories yields an inherently unstable position.

4.2 FRIEDMAN'S OBSERVATION

The preceding objection to constructive empiricism reveals a fundamental point of similarity between van Fraassen's and Carnap's understandings of the truth of theories which postulate unobservable entities. Friedman has observed that despite this similarity, the possibility of transforming an empirically adequate theory into a true one does not pose the same difficulty for Carnap's philosophical neutrality that it does for van Fraassen's evaluation of realism.

As I noted earlier, the semantic view of theories was developed in explicit opposition to the classical logical-empiricist approach of Carnap and others. One of the salient differences between the two approaches is that according to the semantic conception, the observable–unobservable distinction is not a distinction in vocabulary, but a distinction among entities. It is nevertheless reasonable to assume that Carnap's distinction between primitive vocabulary items can be drawn to *reflect* van Fraassen's distinction between entities, so that primitive theoretical terms refer to unobservable entities and the relations between them, while primitive observation terms refer to observable entities and the relations between them. A *relation* is understood

---

Here L_o and L_t are first-order languages with equality all of whose vocabulary is, respectively, observational and theoretical. $L_o \cup L_t$ is the language generated by the observational and theoretical vocabulary of L_o and L_t.

to be observable if the terms it relates are observable but is unobservable otherwise; this terminology is extended in the obvious way to *observation* and *theoretical* relation *symbols*.

According to Carnap, a theory is a class of statements of three types: observation statements, which are composed exclusively of observation terms; theoretical statements, composed exclusively of theoretical terms; and mixed statements which involve both theoretical and observation terms. Carnap's *Ramsey-sentence reconstruction* of theories amounts to the claim that a theory's *factual content* is expressed by its *Ramsey sentence*, where this is the sentence which results when, in the conjunction of the theory's theoretical, observational, and mixed postulates, theoretical terms are replaced by new existentially bound variables of the appropriate logical category. The Ramsey sentence is naturally read as asserting that *there are theoretical relations* that form the basis of an interpretation under which the matrix of the sentence[6] is satisfied, and in particular under which all the observational consequences of the theory hold. Hence if one grants that the distinction between observable and unobservable entities can be reflected by Carnap's distinction between primitive observational and theoretical vocabulary, it is easily seen that the claim that *there is some model of a theory* into which it is possible to fit the observable phenomena countenanced by the theory is precisely what is asserted by the claim that the Ramsey sentence of the theory is true.

The difference, according to Friedman, between Carnap's empiricism and van Fraassen's constructive empiricism is that for Carnap, not only does the Ramsey sentence capture the entire factual content of the theory, but this factual content is not based on anything more than the theory's empirical adequacy: for Carnap, to assert that a theory is true – to assert that the content of the theory is indeed *factual* – is precisely the same as asserting that the theory is empirically adequate. In other words, Carnap holds that there is no gap between an empirically adequate theory and a true one. Hence the observation that played such a fundamental role in our discussion of van Fraassen's defense of constructive empiricism – the observation that on van Fraassen's understanding of how the notions of truth and empirical adequacy are applied to theories, an empirically adequate theory can always be transformed into a true one – does not undermine any thesis of Carnap's. In particular, this observation cannot be used to undermine Carnap's claim to have advanced a view that is *neutral* on

---

[6] By the *matrix* of a Ramsey sentence I mean the formula that results when the "new" existential quantifiers are deleted.

the question of realism, as it did van Fraassen's claim that constructive empiricism is justifiably *agnostic* about a question to which realism is unjustifiably *committed*. Indeed, it makes little sense to be agnostic about truth but committed to empirical adequacy if the condition for a theory's being empirically adequate can be transformed into the condition for its being true. However, under these circumstances it *is* reasonable to maintain a position of neutrality concerning the proposition that the possession of one of these attributes is preferable to the possession of the other.

4.3 CARNAP'S NEUTRALITY

Friedman's analysis suggests that Carnap's position is dialectically superior to that of constructive empiricism. But I will now argue that there is a subtler but equally telling objection to Carnap's reconstruction of theories and the account of realism that emerges from the approach to the controversy recorded in his *Introduction to the Philosophy of Science* and elaborated in his "Replies and systematic expositions."[7]

Let me begin by recalling an abstract example of a theory that reveals an important distinction among theories that in Carnap's reconstruction count as empirically adequate.[8] The language of the theory that comprises this example has just one observation term and one theoretical term: the former is the binary relation symbol R and the latter is the unary relation symbol A. The first-order languages with identity that are defined over the nonlogical vocabularies {R} and {A} will be denoted by L_o and L_t, respectively; $L_o \cup L_t$ denotes the language generated by {A, R}.[9] (Notice therefore that $L_o \cup L_t$ includes formulas that contain *both* observation and theoretical vocabulary items, as well as formulas of L_o and L_t.)

---

[7] Carnap (1966 and 1995) and Carnap (1963b), respectively.

[8] See van Benthem (1978) for a discussion of this theory and related extensions of it, as well as Chapter 7 of this volume. The example goes back to Swijtink (1976). (A very similar case was discovered by Putnam.) The point it establishes – that there exist theories all of whose observational consequences are true, but which are not themselves true in any expansion of a model of their observational consequences – was first noted by William Craig. Craig never published his discovery, and Putnam has only recently done so; see Putnam (2012). That Carnap was aware of Craig's example is clear from a fragment in the Hillman Library Archives of Scientific Philosophy at the University of Pittsburgh. The fragment (file no. RC 088–19–02) is entitled "Notes given by William Craig, February 25, 1961" and is annotated in Carnap's handwriting. I do not know whether Carnap's views were in any way influenced by Craig's example.

[9] I sometimes use T for the example I am about to describe, and sometimes for an arbitrary theory; a similar remark applies to my use of the terms L_o, L_t, and $L_o \cup L_t$. The context will always make clear which is intended.

Let φ be an L_o-sentence which axiomatizes the complete theory of the natural numbers with $<$. Then the *toy theory* T is the theory in $L_o \cup L_t$ consisting of φ together with sentences expressing the fact that A is non-empty, that the first element (first with respect to $<$, of course) is not in A, and that the complement of A is closed under immediate successors. (This last condition means that if x belongs to the complement of A, then for any y such that xRy, if there is no z such that xRz and zRy, then y belongs to the complement of A.)

The L_o-consequences of T are evidently true in the standard model $\langle N, < \rangle$ of the natural numbers under $<$. But T holds only in nonstandard models, i.e. in models with infinitely large or nonstandard natural numbers. Essentially the same is true of T's Ramsey sentence, $\exists XT(R, X)$, even though it is strictly weaker than T. This is because on any interpretation of L_o the complement of A must contain the first element of the $<$-sequence and must be closed under successors. So consider the standard second-order model $\langle N, \wp(N_i)_{i\in I}, < \rangle$, where for each i, $\wp(N_i)$ is the power set of the i-th Cartesian power of N. In this model the domain over which the second-order variables range is the true power set of the domain of individuals – in the present case, the power set of the set of natural numbers. For the Ramsey sentence of T to hold in $\langle N, \wp(N_i)_{i\in I}, < \rangle$, there must be a set in $\wp(N)$ which contains zero, is closed under successors, and does not contain every element of N. But this is impossible when $\langle N, < \rangle$ is standard.

The theory T is empirically adequate but not true, if by 'true' we mean true in an expansion of the standard model of the deductive observational consequences of T, and if by 'empirically adequate' we mean having only deductive observational consequences that are true in the standard model. In fact T is an example of a theory which is empirically adequate (with respect to the standard interpretation of its observation vocabulary), but not true under any *expansion* of the standard model to a model for the language {R, A}. Since the theory T is not true in *any* such expansion, it follows that its Ramsey sentence is also not true in any such expansion. Hence T is an example of a theory whose Ramsey sentence is not true in any expansion of the standard model of T's observational consequences to a model for a language based on T's observation and theoretical vocabulary.

For Carnap the controversy between realists and instrumentalists over theories that postulate theoretical entities *purports* to raise a genuine question about the existence or reality of a kind of entity, but like some other philosophical controversies of this character, it should rather be understood as raising a question about the use of a preferred vocabulary – in the present case, the vocabulary of theoretical terms. Carnap holds that realists are inclined to endorse their use, while instrumentalists eschew their use:

[T]he question should not be discussed in the form: 'Are theoretical entities real?' but rather in the form: 'Shall we prefer a language of physics (and of science in general) that contains theoretical terms, or a language without such terms?' From this point of view the question becomes one of preference and practical decision. (Carnap 1995, p. 256)

The controversy therefore concerns the practical advantages of a reconstruction of the language of a theory that retains its theoretical vocabulary over a reconstruction that does not. Since the question is one of language choice, the only reasonable position to adopt is one of tolerance, thus allowing that there might be practical advantages to the pursuit of both types of reconstruction depending on the purposes the reconstruction is intended to serve. The correct view of the realism–instrumentalism controversy is one that recognizes it to concern the practical advantages and disadvantages that accrue to the use of theoretical vocabulary while maintaining a position of neutrality regarding the existence questions the controversy is alleged to raise.

The fact that there are theories which are not true in any expansion of the standard model of their true observation sentences bears on Carnap's view of realism and his neutrality regarding the realism–instrumentalism controversy. For if T is a theory that is not true in a mere expansion of the standard model of its observational consequences, it cannot be maintained that the difference between realists and instrumentalists is merely a difference in their preference for theoretical language. One of the lessons of our discussion of constructive empiricism was that even if a theory cannot be modeled in an *expansion* of the standard model of the observation sentences, it can be modeled in an *extension* of that model. But modeling a theory in an extension of a model of its true observational consequences is incompatible with Carnap's claim that the difference between realists and instrumentalists consists in *nothing more* than their willingness or unwillingness to use theoretical vocabulary. If, in order to model a theory, it is necessary to interpret its theoretical vocabulary in an extension of the model of its true observation sentences, then the difference between realism and instrumentalism will sometimes rest on more than a decision about language. Whether it does will depend on the nature of the theory under discussion.

The abstract example of our toy theory exploits the distinction between a standard and a nonstandard interpretation of the toy theory's "observation language." In order to apply this discussion to theories which are not merely abstract possibilities, it is necessary to suppose that the notion of a standard interpretation can be extended to the observation languages that arise in the context of a Carnapian reconstruction of the language of science, and it is necessary to suppose that such a reconstruction will grant that observation languages should be given their standard interpretation. Under these conditions, Carnap's explication of the difference between realism and instrumentalism elides the difference between two types of theory and their associated Ramsey sentences. There are theories whose Ramsey sentences are true in an *expansion* of the standard model of the theory's observational consequences, and there are theories whose Ramsey sentences are true only in an *extension* of such a model. The former type of theory supports an interpretation of the realism–instrumentalism controversy along the lines Carnap proposes as a mere difference in preferred vocabulary. For theories of this type, we need only to broaden the domain of the interpretation function to include theoretical terms. But in the case of a theory that is not true in such an expansion – and for which it is necessary to modify the *structure* to which the interpretation function maps theoretical terms – the provision of a model requires more than a mere change of language. Hence the difference between realism and instrumentalism is merely linguistic in the sense Carnap requires only if the theories under consideration are of a certain kind – only if they are true in an expansion of the standard model of their true observation sentences. It follows that Carnap cannot claim to have given a *general* account of the dispute between realists and instrumentalists; so far as this controversy is concerned, it may even be that *all* the interesting cases demand an appeal to extensions of models, and therefore fall outside the purview of Carnap's characterization.

To summarize our discussion of Carnap on realism and the reconstruction of theories, the Ramsey sentence might share with a theory the property that it does not hold in any expansion of a model of the true observation sentences that are formulable in the observation language of the theory. This will be the case when the models of the theory's true observation sentences are standard but the theoretical sentences introduce nonstandard elements. Under these circumstances, we must move to an extension of a model of the true observation sentences to model the theory, and this is more than a mere change of language. But for Carnap the only model-theoretic difference that should separate realists from instrumentalists is the difference between a model of the true observation sentences and an expansion of it to one that interprets the theoretical vocabulary as well as the observation vocabulary. In this case, their difference is reasonably described as merely linguistic. But if a theory can hold only in an extension of the model of its true observation sentences, the difference will also rest on the character of the domain of the model, and this is more than a mere difference in attitude toward the theoretical language.

The foregoing discussion of Carnap's analysis of realism allows us to make sense of an otherwise puzzling aspect of his later thought by suggesting a rationale for how he could continue to maintain, in the face of the considerations we have just raised, that the dispute between realists and instrumentalists should be represented as turning entirely on a question of language choice. The puzzling aspect I am referring to is Carnap's invocation of an arithmetical interpretation of the Ramsey sentence. In his "Replies and systematic expositions," Carnap argues that while the Ramsey sentence

> does indeed refer to theoretical entities by the use of abstract variables . . . *it should be noted that these entities are . . . purely logico-mathematical entities*, e.g., natural numbers, classes of such, classes of classes, etc. Nevertheless, [the Ramsey sentence] is obviously a factual sentence. It says that the observable events in the world are such that there are numbers, classes of such, etc., which are correlated with the events in a prescribed way and which have among themselves certain relations; and this assertion is clearly a factual statement about the world. (Carnap 1963b, p. 963, emphasis added)

On the assumption that the "reality" of mathematical objects is a matter of framework choice on which both realists and instrumentalists might very well agree, Carnap can argue that there is no significant difference between an appeal to an extension and an appeal to an expansion of a model of the theory's true observation sentences, since the domain of an extension can always be comprised in part by "objects" whose availability is guaranteed by a choice of language in the broader sense of a choice of linguistic framework. The distinction between theories which arises from the consideration of extensions and expansions merely reveals the role of language choice at the level of the "Platonistic or mathematically extended language." So long as a theory is reconstructed in the Platonistic or mathematically extended language, it will always be possible to find an extension of the standard model of the true observation sentences that is a model of the Ramsey sentence of the theory. The existence of such an extension will be guaranteed by a feature of the linguistic framework to which both realists and instrumentalists are committed. All that will separate the two positions is the decision whether to exploit the extension for the interpretation of the theoretical vocabulary.

Even if the rationale we have just outlined is a natural application of Carnap's views on linguistic frameworks, it has difficulties of its own. Carnap's claim that the Ramsey sentence refers to numbers, classes of them, . . ., classes of . . . classes of them, . . . simply takes for granted that a special approach to theories which introduce unobservables is justified. And indeed, the traditional controversy between realists and instrumentalists is puzzling if one ignores the fact that it has always engaged the status of theories that postulate unobservable entities which it has assumed are justifiably regarded as problematic precisely because they are unobservable. It should be noted that in this respect, constructive empiricism is no different than its logical-empiricist antecedents. Given the pivotal role of this assumption about the status of theories that postulate unobservable entities, any discussion of the debate between realists and instrumentalists that represents it as focused on the question whether theories are "merely instruments for predicting phenomena" is potentially misleading. The use of 'phenomena' may suggest the traditional theory–observation distinction, but what counts as a phenomenon need not be tied to this distinction. And if it is not, there will be no difference that makes a difference between "realist" and "instrumentalist" answers to the question whether theories are mere instruments of prediction.[10]

Carnap's neutrality on metaphysical questions of ontology was given its fullest expression in "Empiricism, semantics and ontology."[11] While Carnap's neutrality is in general to be applauded, the mere fact that the things with which a theory is concerned are unobservable is hardly sufficient to make the question of their reality a metaphysical one and therefore comparable, for example, to the question whether the basic constituents of the world are "continuants" or "events." The examples Carnap singles out for special emphasis as examples of "external" or metaphysical existence questions come from classical philosophical debates over the reality of abstract objects – debates over the reality of properties, numbers, and propositions. The context of the realism–instrumentalism controversy is different, involving as it does molecular and atomic reality. This case is both paradigmatic and typical for our understanding of the controversy. That the debate over atomism – and therefore also the question at issue in the realism–instrumentalism controversy – is not primarily a metaphysical one is borne out by the fact that what one finds in the writings of the debate's most thoughtful participants is not a blanket rejection of atoms and molecules on the ground that they are unobservable, but a concern that the theories on offer are able to claim for their support only a kind of inference to the best explanation.

---

[10] For an illuminating discussion of the controversy along these lines, see Stein (1989).
[11] Carnap (1950).

In the case of the atomic hypothesis, realism appears completely vindicated – as even Poincaré conceded a century ago – by the ingenious application of our established methods of proof and evidence. But Poincaré did not always accept the reality of atoms. In his essay "Hypotheses in physics" – which was written before Jean Perrin's investigations into Brownian motion and the nature of "molecular reality" – Poincaré distinguished between hypotheses which are "natural and necessary," such as the suppositions "that the influence of very distant bodies is quite negligible, that small movements obey a linear law, and that [an] effect is a continuous function of its cause," and those that should be qualified as "neutral." He then argued that the atomic hypothesis falls into the second category on the ground that:

In most questions the analyst assumes, at the beginning of his calculations, either that matter is continuous, or on the contrary, that it is formed of atoms. He might have made the opposite assumption without changing his results. He would only have had more trouble to obtain them – that is all.

Regarding an analyst whose calculations assume the atomic hypothesis, Poincaré continues,

If . . . experiment confirms [such an analyst's] conclusions, will he suppose that he has proved, for example, the real existence of atoms?[12]

The clear suggestion of "Hypotheses in physics" is that no respectable analyst would make such a supposition because he would recognize that the hypothesis has the epistemic status of a convenient aid to calculation.

However, in a lecture delivered near the end of his life,[13] Poincaré argued that the atomic hypothesis is decidedly *not* neutral.[14] There Poincaré remarks that for a long time the hypothesis was at best a convenience because the question of its truth was not susceptible to resolution. He goes on to say that more recently advances of importance have occurred that allow us to transcend this situation, so that we can now confidently assert that we have abundant proof – in the form of Perrin's results – that the atomic hypothesis is correct. This is established by an appeal to the variety of independent sources of evidence for the hypothesis:

[12] All quotations in this paragraph are from Poincaré (1900, p. 135).
[13] The lecture occurred on March 7, 1912, and was later published as "Les conceptions nouvelles de la matière" (Poincaré 1912a). See Demopoulos, Frappier and Bub (2012) for a translation and discussion of this essay; all quotations are from this translation.
[14] This is not to say that it is "necessary," since Poincaré's distinction does not purport to *partition* the class of all possible hypotheses.

More than a dozen entirely independent processes, that I would not be able to enumerate without tiring you, lead us to the same result. If there were more or fewer molecules per gram, the brightness of the blue sky would be entirely different; incandescent bodies would radiate more or radiate less, and so on. (Poincaré 1912a)

It is not merely that we can give a "just so" story about, for example, the blue of the sky given the particulate structure of the atmosphere, but that the value of Avogadro's number – which is established in a variety of ways, all of them independent of the consideration of this particular question – mandates this color rather than another. Poincaré concludes saying that "There is no denying it, we see atoms" – by which I take him to mean that the weight and quality of the evidence is overwhelmingly in their favor.

The point is reiterated – and the general methodological principle on which it is based is formulated – in a lecture given shortly afterward when, discussing different kinds of confirmation which may attach to a hypothesis, Poincaré raises the case where the novelty of a phenomenon is only apparent because its connection with those for which the hypothesis was originally adduced is so close that any hypothesis which accounts for the one must necessarily account for the other. This situation is contrasted with a case of genuine novelty, not in a psychological sense of 'novelty,' which Poincaré has just explained is not at issue, but in the sense of "a coincidence which could have been anticipated and could not be due to chance." Such coincidences are methodologically decisive, "particularly when a numerical coincidence is involved." Referring to Perrin's results, Poincaré concludes: "Now, coincidences of this type have, in recent times, confirmed the atomistic concepts."[15]

In connection with Poincaré's understanding of the significance of Perrin's discoveries, it is interesting to consider the recent assessments by Penelope Maddy and Bas van Fraassen. Maddy (2001, p. 59) holds that prior to Perrin's results, the atomic hypothesis was regarded as empirically adequate. Poincaré is pointing out that this is radically mistaken, and that the evidence Perrin brought to bear on the question was decisive in establishing for the first time the empirical adequacy of the hypothesis of "molecular reality." According to Poincaré, Perrin's results established the hypothesis so conclusively that the reality of atoms and molecules could no longer be reasonably doubted. Moreover this contention of Poincaré's does

---

[15] Poincaré (1912b, p. 90), which is based on a lecture of April 11, 1912. Poincaré was remarkably perceptive concerning the methodological situation. For a recent discussion of Perrin's achievement along the lines of Poincaré's remarks, together with a highly accessible account of the chemical and physical tradition to which Perrin contributed, see Chalmers (2011).

not depend on his having adopted a weaker notion of truth, or on the assumption that the hypothesis of molecular reality has been reduced to an assertion about "phenomena." It depends on the fact that the theoretical and experimental considerations that accompanied Perrin's demonstration of empirical adequacy conform to the relevant canons of evidence in a way that earlier attempted demonstrations did not, and it is for this reason that Perrin's work established the reality of atoms and molecules.

Although van Fraassen effectively concedes Poincaré's account of the significance of Perrin's work, he argues that Perrin failed to address anything that might be called the question of *realism*. For van Fraassen the issue of realism could be settled by "the empirical grounding" of the kinetic theory only if the problem concerned the status of unobservables, and then only if this were the problem Perrin actually addressed. Since the scientific tradition did not question molecular reality on the basis of a general skepticism about unobservable entities, van Fraassen concludes that it is inappropriate to suppose that results like Perrin's bear on realism. But as we have seen: (i) the demonstration of the atomic hypothesis shows that the philosophical problem of realism versus instrumentalism as the positivist and neo-positivist philosophical tradition conceived it – namely as a problem about the reality of theoretical, i.e. unobservable, entities – *is* susceptible to a definitive resolution in one important case; and since this is sufficient to show that there cannot be a *general* problem with a class of entities that derives from the mere fact of their unobservability, it evidently *does* bear on the philosophical debate; (ii) the belief to the contrary is plausible only so long as the methodological situation has not been properly grasped; and (iii) if there is a purely philosophical or metaphysical problem of realism versus instrumentalism, its source must lie elsewhere than with the status of hypotheses involving unobservables.

Returning to Carnap's proposal for how the Ramsey sentence should be understood, we may conclude that in so far as the realism–instrumentalism controversy fails to pose a *metaphysical* problem of the existence of unobservable entities, there is no basis for Carnap's interpretation of the Ramsey sentence as a statement about purely logico-mathematical entities. The most one can say on behalf of such an interpretation is that it supports a position of neutrality on a question which Carnap and many others[16] radically misunderstood, a question which is susceptible to a definitive answer.

---

[16] Here Nagel (1961) comes to mind; see especially Chapter 6, "The cognitive status of theories," to which Carnap refers (1966, p. 256, fn). In the second edition of Carnap (1966), the reference to Nagel is replaced by the citation of Carnap (1950).

4.4 THE ROLE OF METAPHYSICS IN THE REALISM–INSTRUMENTALISM CONTROVERSY

Is there, then, *no* metaphysical question that can reasonably be associated with the realism–instrumentalism controversy? Certainly not: the *metaphysical* issue that separates realists from instrumentalists is not the reality of unobservables – in all cases of interest, if the question of their reality can be settled at all, it can be settled by standard methods – but a disagreement over the concept of truth that is appropriate to empirical theories which exhibit the abstractness and generality that is characteristic of modern physics. To address this issue, I propose to formulate two frameworks – one "realist," the other "anti-realist" – for understanding the semantic properties of theories. I will argue that the two frameworks are *not* on a par, and that the realist framework is preferable to its anti-realist alternative.

The theories that motivated Carnap's reconstruction are distinguished by their mathematical sophistication and their amenability to being represented in a language which distinguishes between observational and theoretical vocabulary. The Ramsey-sentence explication of the factual content of these theories makes Carnap's conception of their truth particularly clear. The central assumption of the explication that bears on this issue is expressed by what I will call *the structuralist thesis*, namely the thesis that the evaluation of the truth of a statement which contains a theoretical term depends only on the term's logical category.[17]

The structuralist thesis is a simple extrapolation from Hilbert's understanding of the essential *generality* of mathematical theories. Hilbert argued that the proper formulation of a mathematical theory should not be constrained by the demand that it preserve preconceptions regarding the nature of the theory's primitive notions. Otherwise a theory's applications – both its internal or mathematical applications and its applications outside of mathematics – would be unduly limited, and the theory's generality thereby compromised. But the application of the structuralist thesis to physical theories lacks the straightforward motivation that informed Hilbert's goal of capturing the generality of mathematics. This goal is naturally met by the idea that the terms of a mathematical theory impose no conditions on their interpretation other than what is demanded by the notion of a nonlogical constant – which is to say that their interpretation is only required to respect the logical category of the theory's primitives. But the claim that physical theories exhibit the same kind of generality as mathematical theories needs a

---

[17] The structuralist thesis is the central component of the views considered in Chapter 7.

separate argument.[18] To my knowledge, no such argument has ever been given. It has simply been taken for granted that physical theories inherit the generality of the mathematical theories used in their formulation.

The problem that proponents of the structuralist thesis *have* sought to address is how to differentiate the theoretical claims of a physical theory from the statements of pure mathematics. According to Carnap, the distinction rests on the fact that theoretical claims are endowed with *empirical* content while the statements of pure mathematics are not. The empirical content of physical theories is in turn explained as the product of three characteristic features of their preferred reconstruction: (i) the reconstruction of the language of physical theory is based on the distinction between observational and theoretical terms; (ii) the theories of interest include both types of term; and (iii) in addition to observational and theoretical statements, there are, among the statements that comprise a theory's reconstruction, some that are mixed in the vocabulary they contain; these latter statements are the so-called "correspondence rules." Together (i), (ii), and (iii) form the basis for the reconstruction's account of the fact that physical theories, unlike the theories of pure mathematics, have empirical content.

But representing empirical content by the distinction between observation and theoretical terms, and the provision of statements expressing the connections between them, is too coarse a grid to capture the difference between a case in which a hypothesis, like the atomic hypothesis, is formulated so that it is susceptible to definitive resolution by empirical and analytic methods and a case in which its formulation allows only for a justification that is merely hypothetico-deductive. This is a difference that is standardly understood as a difference in empirical content, and it is one that should be reflected by the notion's explication. The fact that it is not reflected in an explication like Carnap's is the source of the positivist and neo-positivist misconception that realists and instrumentalists are divided over a metaphysical existence question: having an insufficiently refined analysis of empirical content, differences that are empirically resolvable are misinterpreted as indicative of opposing metaphysical preconceptions; as a consequence, a substantive question is relegated to a conventional decision about the use of theoretical vocabulary.[19]

---

[18] As Hilbert's correspondence with Frege amply demonstrates, even in mathematics the nature and extent of the generality of its theories was not always clear.

[19] Or, as in a recent proposal of van Fraassen's, the problem is relegated to a hermeneutical question about how best to interpret the meta-theoretical claims of scientists. See van Fraassen (2009, pp. 7–8).

The structuralist thesis blurs the distinction between the mere *satisfiability* of a theoretical claim and its *truth*. This follows from the fact that the thesis equates the truth of a theoretical claim under an arbitrary interpretation – even an "abstract" or mathematical one – with its truth under the claim's standard interpretation. Such a conflation will appear justified in the special case of theoretical claims if one assumes that Carnap's explication of empirical content is correct and complete. For it can then be argued that, having captured the notion of empirical content appropriate to theoretical statements, nothing more than satisfiability can reasonably be demanded of an account of their truth. But since we have independent grounds for believing this explication of empirical content to be inadequate, we have reason to reject any such justification of the structuralist thesis and the view of the truth of theoretical claims to which it leads.

Within the class of empirical theories, our interest is in how a reconstruction analyzes the truth of the statements to which it assigns empirical content. I will call a theory of the truth of such statements a *realist* theory if it holds that the notion of truth is univocal across empirical statements – that is, if, among the statements it counts as having empirical content, the theory makes no distinction in the notion of truth that is applicable to them. I will call a theory *anti-realist* if it isolates a subclass of empirical statements – the *theoretical* statements – and holds that the notion of truth that applies to the statements of this class is strictly weaker than the notion of truth that applies to statements that fall outside the class of theoretical statements. Finally, a *framework* for discussing theories is realist or anti-realist according to whether the theory of truth it incorporates is a realist or anti-realist theory. Clearly, this characterization is very general and allows for realist and anti-realist theories that may not be of much interest. The cases that are of interest to us are those for which the subclass of exceptional statements is singled out on the basis of the observation–theoretical distinction, and the weakening of the notion of truth is the weakening mandated by the structuralist thesis.[20]

On this taxonomy of frameworks for the reconstruction of empirical theories, Carnap's framework is one that favors anti-realism, while the one to which we have been led favors realism. But neither the assumption of the

---

[20] For Carnap the structuralist thesis is expressed by the Ramsey sentence and its use of variables, while for constructive empiricism and the semantic view of theories the thesis is expressed by a model-theoretic claim employing nonlogical constants. There is an important methodological distinction between variables and nonlogical constants that is reflected in the close association of the latter notion with the idea of a *variable domain of interpretation*. But this is not a difference that affects the relevant respect in which constructive empiricism agrees with Carnap's analysis. This agreement is captured by their common conception of the truth of theories involving unobservables; it matters little for the point at issue whether this is formulated model-theoretically or with the notion of a Ramsey sentence.

correctness of realism as opposed to instrumentalism, nor the presumption in favor of a realist conception of the truth of theories that postulate unobservables, is in conflict with Carnap's ontological neutrality.[21] Hence, it cannot be objected that our analysis begs the question against Carnap by assuming the legitimacy of questions he would reject. The first assumption – which expresses our endorsement of realism over instrumentalism in the controversy over atomism – is not question-begging because it does not pertain to a metaphysical question of any kind; and the second assumption – regarding the notion of truth appropriate to theories – does not pertain to a metaphysical question of ontology of the kind Carnap was concerned to reject. There is therefore nothing in the preceding discussion of Carnap's Ramsey-sentence reconstruction of theories that either suggests or requires a reconsideration of Carnap's proposed neutrality on traditional metaphysical questions.

Nor is there anything in the preceding discussion that could be taken as a challenge to what Friedman has observed is a "structuralist strain" in Carnap's later thought. Understood as the claim that physics has a great deal to tell us about the structural properties of the world, structuralism is not committed to the structuralist thesis, but is in the same position as realism in the context of the realism–instrumentalism debate. From our perspective, realism about structure differs from realism only in its focus on abstract properties; as such it forms part of the generally accepted framework of science, and is hardly a matter for serious philosophical disagreement. Our critical focus has been directed at the structuralist thesis and at showing that this is not a thesis we have to accept in order to appreciate the importance of knowledge of structure or to sustain Carnap's neutrality on metaphysical questions of ontology.

APPENDIX: EMPIRICIST STRUCTURALISM

In his recent book, *Scientific Representation: Paradoxes of Perspective*,[22] Bas van Fraassen develops a position – "empiricist structuralism" – which promises to address certain difficulties that his earlier constructive empiricism encountered. In my view, the source of these difficulties is constructive empiricism's contention that the difference it articulates between an empirically adequate theory and a true one can form a compelling basis for agnosticism about theories which postulate unobservable entities.

---

[21] In this respect, the present analysis stands in marked contrast with the view of Maddy (2008) which rejects Carnap's notion of an external question and takes this rejection to undermine his philosophical neutrality. Maddy's views are discussed in Chapter 3.

[22] Van Fraassen (2008).

Recall that constructive empiricism's account of the relation between truth and empirical adequacy is very simple: empirical adequacy is concerned with what a theory says about observable phenomena, while truth is concerned with what a theory says about observable and unobservable phenomena. The key point to notice is that by the semantic view of theories – which is the framework within which constructive empiricism is formulated – both truth and empirical adequacy are analyzed in terms of isomorphism of models. A theory is *empirically adequate* if it contains a model into which the observable phenomena can be isomorphically imbedded. The notion that a theory is true is the simple extension of this idea from observable phenomena to all phenomena, so that a theory is *true* if it contains a model which is isomorphic to the observable and unobservable phenomena. Empirical adequacy thus involves an imbedding or *partial* isomorphism, while truth involves the isomorphism of a model belonging to the theory with the *whole* domain of observable and unobservable phenomena.

We exposed the vacuity of this notion of a true theory using Putnam's model-theoretic argument. But even constructive empiricism's account of an empirically adequate theory is vacuous, since it also rests on the oversight of seeking to explain a notion in terms of isomorphism while leaving the relations which are to be preserved by the isomorphism indefinite and unspecified. As a consequence, both the truth and empirical adequacy of a theory are reduced to something only slightly more than a question of logic. Hence the difficulty Putnam's argument poses for constructive empiricism cuts deeper than the doctrine's account of the truth of theories which postulate unobservable entities, and threatens even the coherence of its analysis of what it is for such a theory to be empirically adequate. In order to rectify the deficiencies of constructive empiricism, empiricist structuralism needs to give a nonvacuous formulation of the claim that a theory is empirically adequate when it contains a model which represents the observable phenomena, and it needs to formulate a coherent rationale for agnosticism about the representation of unobservable phenomena – one that sets such a representation apart from the representation of phenomena.

Scientific Representation articulates several ideas that bear on these two tasks. Regarding the formulation of empirical adequacy, van Fraassen argues first that in the context of its use as a representation, an isomorphism is not simply a relation between two structures, but involves the rational agent in an essential way. Secondly, whether or not it plays a representational role, an isomorphism is not a relation between the world or the observable part of the world and a model, but a relation that concerns two mathematical objects, one of which is a model belonging to the theory, while the other is a model which comprises the data we hope to explain. Thus to say that a theory is empirically adequate is to say that it contains a model into which the data – as identified by such a model – can be isomorphically imbedded, with the tacit understanding that since this is a representational use of the notion of isomorphism, it must implicitly involve the "perspective" of the rational agent to which we just called attention. The opposing view – which van Fraassen rejects – is one that seeks to explain empirical adequacy using only a simple relation between a model belonging to the theory and reality, where, as van Fraassen expounds the view, reality is understood to be "carved at its joints," and is capable of forming a model without reference to the perspective of a rational agent.

Unlike constructive empiricism, empiricist structuralism's agnosticism is not based on the distinction between our acceptance of a theory as empirically adequate and our belief that it is true. For an empiricist structuralist theories do not purport to be "true of reality." Those parts of models which we naively take to represent unobservable phenomena are merely abstract structural schemes by which we anticipate the unfolding of the data demanding explanation. The rational agent is "oriented" and has a "perspective" toward the observable phenomena or data, and this is alleged to undermine the charge of vacuity that it was possible to urge against constructive empiricism's notion of empirical adequacy. So far as the theoretical component of a model is concerned, such a perspective is neither possible nor necessary, since what the model conveys about structure exhausts what our understanding of it requires in order for it to guide us in our anticipation of observable phenomena.

It is not clear that empiricist structuralism is fundamentally different from Carnap's mature rational reconstruction of theories. A Carnapian formulation of the position would run as follows. In the case of observable phenomena, we are in possession of an *interpretation* of our statements about them; having such an interpretation goes beyond purely structural knowledge of the referents of observation terms. But we lack a comparable interpretation of our theoretical claims that takes them beyond the purely structural knowledge they convey; in the case of theoretical claims we must settle for an agnosticism about their subject matter. It follows that our knowledge of the theoretical parts of science – those that, as we commonly say, concern unobservable entities – is in a different position than our knowledge of observable phenomena. So formulated, empiricist structuralism is susceptible to the same objections we urged against Carnap's Ramsey-sentence reconstruction of theories.

This conclusion might be challenged on the ground that the foregoing presentation of empiricist structuralism is at variance with van Fraassen's perception of the problem lying at the basis of *constructive* empiricism that he takes empiricist structuralism to resolve. According to van Fraassen:

The basic perplexity emerges in its purest form when we ask [what it means] to embed the phenomena in an abstract structure. Or to represent them by doing so?[23] . . . Hence the most fundamental question is this: How can an abstract entity, such as a mathematical structure, represent something that is not abstract, something in nature?[24]

Following Reichenbach and others, van Fraassen argues that by speaking of an isomorphic mapping which imbeds the phenomena in an abstract structure we commit ourselves to two *sets* of elements, one for the domain, and another for the codomain of the mapping. Van Fraassen and Reichenbach conclude from this that the use of the notion of an isomorphism to explain the mathematical representation of reality *already* invokes a mathematical representation of the things the mapping correlates. Van Fraassen explains that

Reichenbach was imagining, and discounting, the following naive sort of reply [to the question, 'How can an abstract entity, such as a mathematical structure, represent something that is not abstract, something in nature?']: *what is called for is simply a function, a mapping, between mathematical and physical objects or processes – what is puzzling about that?* The reason he discounts [this reply] is because to define a function we need to have the domain and range defined first – and the question at issue was precisely how that can be done without presupposing that we already have a physical – mathematical relation at hand.[25]

Van Fraassen later returns to this point, remarking that:

we have no difficulty with mappings between two sets of mathematical objects. But notice: to define such a mapping we must identify its domain and its range, plus the relation that effects the coordination between them . . . So here is the problem baldly stated. If the target is not a mathematical object then we do not have a well-defined range for the function, so how can we speak of an embedding or isomorphism or homomorphism or whatever between the target and some mathematical object?[26]

---

[23] In an endnote to this passage, van Fraassen remarks: "The semantic view of theories runs into severe difficulties if these notions are construed either naively, in a metaphysical way, or too closely on the pattern of the earlier syntactic view. Constructive empiricism, which was formulated in the context of the semantic view of theories, shares this vulnerability" (2008, pp. 385–6).

[24] *Ibid.*, p. 240.   [25] *Ibid.*, p. 120.

[26] *Ibid.*, p. 241. An endnote to this passage tells us that "in *The Scientific Image* constructive empiricism was presented in the framework of the semantic view of theories, but seemingly in the shape of the above 'offhand' realist response. See for instance Chapter 3 section 9, p. 64 where I define empirical adequacy using unquestioningly the idea that concrete observable entities (the appearances or phenomena) can be isomorphic to abstract ones (substructures of models). Demopoulos 2003a comments on this passage." I hope it is clear that none of my criticisms of constructive empiricism rest on the charge that it is implicitly committed to the "offhand realist response," still less to realism.

Hence according to van Fraassen and Reichenbach, we cannot, without circularity, use the notion of an isomorphism to explain how phenomena are mathematically represented as "something in nature," since any mapping already assumes a mathematical, and hence abstract, representation of what it acts upon.

But it should be far from obvious that van Fraassen and Reichenbach have succeeded in raising a genuine problem for the representational use of an isomorphism. To begin with, such a use of the notion requires only that we bring the things correlated under a concept; doing so does not by itself constitute a *mathematical* representation. Nor does such a conceptual representation reduce to or presuppose a mathematical representation. Any reasonable formulation of the problem of how mathematics represents reality must be predicated on the assumption that we can provisionally take for granted what is meant by *conceptually representing* reality.

Were we to follow Reichenbach and van Fraassen, we would be forced to reject Frege's celebrated solution to the problem of how arithmetic applies to reality. To see this, recall that for Frege the applicability of arithmetic to reality is explained by the fact that our judgments of cardinality rest on relations between concepts, and concepts sometimes apply to reality. It is a defect of Reichenbach's and van Fraassen's views that they must reject this solution on the ground that the relevant relation between concepts – the relation of one–one correspondence – assumes the mathematical representation of the domain and codomain of the correspondence. On a view like Reichenbach's or van Fraassen's, the Fregean account of the application of *arithmetic* to reality would presuppose an understanding of the application of *mathematics* to reality. But for Frege the representation of the domain and codomain are conceptual, not mathematical: no mathematical notion comes into it. Hence, if we can assume that we understand how concepts apply to reality, the application of arithmetic to reality presents no further difficulty. The same holds of the mathematical representation of reality. The use of the concept of an isomorphism in any such account also rests on the clarity of the representation of things by concepts. As with the Fregean account of application, so also here, the representational use of a correspondence in an account of mathematical representation – even a correspondence that is an isomorphism – does not introduce a circularity into the analysis. I conclude that the problems with which empiricist structuralism has to deal are not so general as van Fraassen understands them to be, but arise from the fact that what constructive empiricism means by the empirical adequacy and truth of theories fails to capture essential components of these notions.

CHAPTER 5

Bertrand Russell's The Analysis of Matter: *its historical context and contemporary interest*

With Michael Friedman

The Analysis of Matter[1] is perhaps best known for marking Russell's rejection of phenomenalism (in both its classical and methodological forms) and his development of a variety of Lockean representationalism – Russell's causal theory of perception. This occupies Part II of the work. Part I, which is certainly less well known, contains many observations on twentieth-century physics. Unfortunately, Russell's discussion of relativity and the foundations of physical geometry is carried out in apparent ignorance of Reichenbach's and Carnap's investigations of the same period. The issue of conventionalism in its then contemporary form is simply not discussed. The only writers of the period who appear to have had any influence on Russell's conception of the philosophical issues raised by relativity were Whitehead and Eddington. Even the work of A. A. Robb fails to receive any extended discussion,[2] although Robb's causal theory is certainly relevant to many of Russell's concerns, especially those voiced in Part III, regarding the construction of points and the topology of space-time. In the case of quantum mechanics, the idiosyncrasy of Russell's selection of topics is more understandable, since the Heisenberg and Schrödinger theories had only just been put forth. Nevertheless, it seems bizarre to a contemporary reader that Russell should have given such emphasis to G. N. Lewis's suggestion that an atom emits light only when there is another atom to receive it – a suggestion reminiscent of Leibniz, and one to which Russell frequently returns.[3] In short, the

---

First published in *Philosophy of Science* 52 (1985): 621–39. An earlier version of this paper, under the title, "Russell's theory of theories," was presented at the Grover Maxwell Memorial Conference on Bertrand Russell's Philosophy of Science, June 1982, with Michael Friedman commenting. The published version of the paper incorporates Friedman's extension of my discussion of Russell and Newman to views held by Schlick, Carnap, and Wittgenstein.

[1] Russell (1927); citations of this work use the acronym *AM*.

[2] See Winnie (1977, p. 157) for a discussion of Russell's remarks on Minkowski space-time and the bearing on them of a better appreciation of causal structure.

[3] See especially *AM* (pp. 125–9).

philosophical problems of modern physics with which Russell deals seem remote from the perspective of post-positivist philosophy of physics. But if the observations on philosophy of physics seem dated, this is not true of the theory of theories that the book develops. As Grover Maxwell emphasized,[4] it is possible to extract from the book a theory of theories that anticipates in several respects the Ramsey-sentence reconstruction of physical theories articulated by Carnap and others many decades later.

5.1

The heart of the theory of *The Analysis of Matter* is the claim that our knowledge of the external world is purely structural. For Russell, this thesis was based on the contention that we are not "directly acquainted" with physical objects. On a more neutral reading, the basis for the thesis is the belief that the reference of the class of theoretical predicates has an explanation in terms of the reference of another family of predicates. Such an explanation, if it could be given, would be highly nontrivial since no definitional or reductive relation between the two classes of predicates is claimed, although there is for Russell something like a reductive relation between our knowledge of theoretical properties and our knowledge of perceptual properties. In this respect the theory of *The Analysis of Matter* stands in marked contrast to the phenomenalism of Russell's 1914 external world program.[5] The point of the latter, of course, is to assume only perceptual objects (sensibilia, or actual and possible sense-data), and perceptual relations (especially perceptual similarity relations), and to *explicitly define* all other objects and relations from these. In *The Analysis of Matter*, Russell wishes to exploit the notion of logical form or structure to introduce scientific objects and relations by means of so-called *axiomatic* or *implicit definitions*. Thus, if we represent a scientific theory by

$$\vartheta(O_1, \ldots, O_m; T_1, \ldots, T_n), \tag{1}$$

where $O_1, \ldots, O_m$ are observational or perceptual terms, and $T_1, \ldots, T_n$ are theoretical terms, then Russell should be prepared to accept the Ramsey sentence

$$\exists X_1 \ldots \exists X_n \, \vartheta(O_1, \ldots, O_m; X_1, \ldots, X_n) \tag{2}$$

---

[4] See especially Maxwell (1968) and (1970).   [5] As articulated in Russell (1914a) and (1914b).

as the proper statement of our scientific knowledge. And (2) is legitimate whether or not the theoretical terms in (1) are explicitly definable from the observational terms; indeed, the whole tenor of Russell's discussion in *The Analysis of Matter* is that theoretical terms will not be explicitly definable in purely observational terms.[6]

It is not entirely clear why Russell abandoned his earlier phenomenalism. His explicit discussion of the issue in Chapter xx of *The Analysis of Matter* is rather inconclusive. Postulating nonperceptible events allows us to maintain a very desirable continuity in stating the laws of nature (*AM*, pp. 216–71), but it appears that "ideal" perceptible events (the sensibilia of 1914) would serve equally well (*AM*, pp. 210 and 213). A perhaps more interesting reason for Russell's shift emerges from his discussion of the then current physics in Part I. Briefly, the point is that a phenomenalist reduction in the style of his program of 1914 does not do justice to the novel "abstractness" of modern physics. Compared with the knowledge expressed in classical physics or common sense, for example, the knowledge conveyed by twentieth-century physics appears to be on a higher level of abstraction; it is "structural" or "mathematical" in an important new sense. For Russell, this leads to a partial skepticism regarding our knowledge of the physical world:

Whatever we infer from perceptions it is only structure that we can validly infer; and structure is what can be expressed by mathematical logic. (*AM*, p. 254)

The only legitimate attitude about the physical world seems to be one of complete agnosticism as regards all but its mathematical properties. (*AM*, p. 270)[7]

The view expressed in these passages is very strongly anticipated in *Introduction to Mathematical Philosophy*,[8] where its Kantian motivation is quite evident:

There has been a great deal of speculation in traditional philosophy which might have been avoided if the importance of structure, and the difficulty of getting

---

[6] Notice that whether the terms are implicitly or explicitly definable depends on the logical framework within which the definitions are to be constructed. Is (1) a first-order theory with identity? An extension of first-order set theory? A formulation in some higher-order logic? These differences matter because, if one is working in a sufficiently strong background language, one can often *transform* axiomatic or implicit definitions into explicit definitions (cf. Quine [1964], for one such strategy). Of course this way of expressing the issue is anachronistic, since these distinctions were not available to Russell. Consequently the distinction between implicit and explicit definition is not so clear-cut for Russell. Thus, for example, Russell appears willing to allow implicit definitions even in 1914 when, in "The relation of sense-data to physics" (1914b), at the conclusion of sect. xi, he defines physical things as "those series of appearances whose matter obeys the laws of physics," though that there are such series is of course, for Russell, an empirical fact.

[7] See also Chapter xiv, especially the discussion of Eddington (*AM*, pp. 136–8).   [8] Russell (1919).

behind it, had been realized. For example, it is often said that space and time are subjective, but they have objective counterparts; or that phenomena are subjective, but are caused by things in themselves, which must have differences *inter se* corresponding with the differences in the phenomena to which they give rise. Where such hypotheses are made, it is generally supposed that we can know very little about the objective counterparts. In actual fact, however, if the hypotheses as stated were correct, the objective counterparts would form a world having the same structure as the phenomenal world, and allowing us to infer from phenomena the truth of all propositions that can be stated in abstract terms and are known to be true of phenomena. If the phenomenal world has three dimensions, so must the world behind phenomena; if the phenomenal world is Euclidean, so must the other be; and so on. In short, every proposition having a communicable significance must be true of both worlds or of neither: the only difference must lie in just that essence of individuality which always eludes words and baffles description, but which, for that very reason, is irrelevant to science. Now the only purpose that philosophers have in view in condemning phenomena is in order to persuade themselves and others that the real world is very different from the world of appearance. We can all sympathize with their wish to prove such a very desirable proposition, but we cannot congratulate them on their success. It is true that many of them do not assert objective counterparts to phenomena, and these escape from the above argument. Those who do assert counterparts are, as a rule, very reticent on the subject, probably because they feel instinctively that, if pursued, it will bring about too much of a *rapprochement* between the real and the phenomenal world. If they were to pursue the topic, they could hardly avoid the conclusions which we have been suggesting. In such ways, as well as in many others, the notion of structure . . . is important. (*AM*, pp. 61–2)

To sum up, on Russell's "structuralism" or "structural realism" of "percepts" we know *both* their quality and structure (where Russell's use of the term 'quality' includes relations), while of external events we know only their structure. The distinction is, in the first instance, between properties and relations of individuals and properties and relations of properties and relations. Structural properties are thus a particular subclass of properties and relations: they are marked by the fact that they are expressible in the language of pure logic. Unlike Locke's distinction between primary and secondary qualities, the structure–quality distinction does not mark a difference in ontological status: external events have both structure and qualities – indeed when we speak of the structure of external events, this is elliptical for the structural properties of their qualities. Neither is it the case that one is "more fundamental" than the other or that qualities are "occurrent" while structure is a power. What is claimed is a deficiency in our knowledge: of external events we know only the structural properties of their properties and relations, but we do not know the properties and relations themselves.

Of course physical knowledge falls within the scope of this claim, so that the theory of perception immediately yields the consequence that physical theories give knowledge of structure and *only* knowledge of structure.

Russell's emphasis on structure has close affinities with much other work of the period, especially with the two classics of early (pre-1930) positivism: Moritz Schlick's *General Theory of Knowledge* (1918/1925) and Rudolf Carnap's *Der logisches Aufbau der Welt* (1928).

The "critical realism" of *General Theory of Knowledge* is identical in almost every respect to the structural realism of *The Analysis of Matter*. Schlick argues that modern physics deals with real unobservable entities (atoms, electrons, the electromagnetic field) which cannot be understood as logical constructions out of sense-data in the manner of Mach's *Analysis of Sensations* or Russell's 1914 external world program (Schlick 1918/1925, sect. 26). Such entities are not experienceable, intuitable, or even picturable; accordingly, Schlick goes so far as to call them "transcendent" entities and "things-in-themselves" (sect. 25). Nevertheless, this "transcendence" presents no obstacle to our knowledge or cognition, for *knowledge relates always to purely formal or structural properties* – not to intuitive qualities or content. Thus, while we cannot experience or intuit the entities of modern physics, we can grasp their formal or structural features by means of axiomatic or implicit definitions in the style of Hilbert's *Foundations of Geometry* – and this is all that knowledge or cognition requires (sects. 5–7). Although the similarities with Russell's view are patent, there is one significant difference: Schlick draws a sharp contrast between knowledge (*Erkennen*) and acquaintance (*Kennen*). On his account *knowledge* by acquaintance is a contradiction in terms, for only structural features are ever knowable (sect. 12). We are acquainted with or experience (*erleben*) some qualities (content), but we have knowledge or cognition of none. As we have seen, for Russell we know both the form and the content of percepts.

Carnap's *Aufbau* does not embrace structural realism. *All* concepts of science are to be explicitly defined within a single "constructional system" whose only nonlogical primitive is a "phenomenalistic" relation R_s of *recollected similarity*. Yet the form–content distinction and the notion of logical structure are equally important. To be sure, all concepts of science are to be constructed from a basis in the given, or "my experience," but the *objectivity* of science is captured in a restriction to purely structural statements about this given basis. Although the "matter" or content of Carnap's constructional system is indeed subjective or "autopsychological" – and therefore private and inexpressible – what is really important is the logical form or structure of the system. For it is logical form and logical form alone that makes objective knowledge and communication possible.[9] As Carnap expresses it:

[S]cience wants to speak about what is objective, and whatever does not belong to the structure but to the material (i.e. anything that can be pointed out in a concrete ostensive definition) is, in the final analysis, subjective. One can easily see that physics is almost altogether desubjectivized, since almost all physical concepts have been transformed into purely structural concepts ... From the point of view of construction theory, this state of affairs is to be described in the following way. The series of experiences is different for each subject. If we want to achieve, in spite of this, agreement in the names for the entities which are constructed on the basis of these experiences, then this cannot be done by reference to the completely divergent content, but only through the formal description of the structure of these entities. (*Aufbau*, sect. 16)[10]

Characteristically, Carnap does not rest content with a simple reference to logical form or structure; he turns his "objectivity requirement" into a definite technical program: the program of defining all scientific concepts in terms of what he calls *purely structural definite descriptions* (*Aufbau*, sects. 11–15). Such a definition explains a particular empirical object as the unique entity satisfying certain purely formal or logical conditions; the *visual field*, for example, is defined as the unique five-dimensional "sense class" (*Aufbau*, sects. 86–91). The point is that purely structural descriptions contain no nonlogical vocabulary; ultimately, we will need only variables and the logical machinery of *Principia Mathematica* (or set theory).[11] Thus, while Russell in *The Analysis of Matter* would formulate our physical knowledge in the manner of (2) (i.e. by turning all theoretical terms into variables), Carnap in the *Aufbau* goes much further: all terms whatsoever are to be replaced by variables. (The reader might very well wonder how Carnap's

---

[9] That is, for Carnap the form–content distinction is drawn in such a way that form is objective and communicable, whereas content is neither. Schlick came to view the form–content distinction in much the same way in his later writings, while Russell came to place much less stress on this (somewhat romantic) view of the distinction. (For Schlick's later view of these matters see Schlick [1926 and 1932].) Schlick explicitly acknowledges the influence of the *Aufbau* and of Wittgenstein's *Tractatus*.

[10] In the explanatory references to this section, Carnap refers to the passage from Russell's *Introduction to Mathematical Philosophy* (pp. 61–2) quoted earlier. Immediately before this section, he refers to the doctrine of implicit definition that Schlick expressed in *General Theory of Knowledge*.

[11] As Carnap himself emphasized, this kind of program for individuating concepts by means of purely formal properties only begins to make sense in the context of modern, *polyadic* logic. Monadic concepts correspond to unstructured sets whose only formal property is their cardinality. Polyadic concepts have such diverse formal properties as transitivity, reflexivity, connectedness, dimensionality, and so on. In Russell's terminology (the terminology of *Introduction to Mathematical Philosophy* and *The Analysis of Matter*) only polyadic concepts have "interesting" relation-numbers.

one nonlogical primitive R_s is to be itself eliminated in favor of variables. We shall return to this later.)

5.2

We believe there are insurmountable difficulties with the theory of theoretical knowledge just outlined. So far as we are aware these difficulties were first raised by M. H. A. Newman in an article published in *Mind* in 1928. Newman's paper – the only philosophical paper he ever published – is not as well known as it deserves to be. (The prose is quite delightful, and for that reason alone, it deserves a wider readership.) This paper contains much that is of interest today, and so we propose to discuss the structuralism of Russell's *The Analysis of Matter* from the perspective of Newman's paper.

Newman begins his discussion with the observation that on any theory of scientific knowledge, the question of the truth of at least some propositions of physics should turn out to be nontrivial. Consider, for example, the question whether matter is atomic. Newman observes that on any account "this is a real question to be answered by consideration of the evidence, not a matter of definition."[12] In fact, whether matter is atomic is a question that concerns the holding of various "structural properties," although they are of a fairly high level of abstraction. (It entails, for example, that physical objects have isolable constituents, and that these constituents have a certain theoretically characterizable autonomy.) Newman's point is that whether the world exhibits such properties is a matter to be discovered, not stipulated, and we may demand of a theory of theories that it preserve this fact. The gist of his criticism of Russell is that (with one exception) Russell's theory does not satisfy this constraint. Let us see how the argument goes.

The difficulty is with the claim that *only* structure is known. On this view, the world consists of objects, forming an aggregate whose structure with regard to a certain relation R is known, say [it has structure] W; but of ... R nothing is known ... but its existence ... all we can say is, *There is* a relation R such that the structure of the external world with reference to R is W. [But] any collection of things can be organized so as to have the structure W provided there are the right number of them. (Newman 1928, p. 144)

Thus, on this view, only cardinality questions are open to discovery. Every other claim about the world that can be known at all can be known a priori

---

[12] Newman (1928, p. 143). All references to Newman are to this article.

as a logical consequence of the existence of a set of α-many objects. For, given a set A of cardinality α, we can, with a minimal amount of set theory or second-order logic, establish the existence of a relation having the structure W provided that W is compatible with the constraint that the cardinality of *A* is α. (The relevant theorem from set theory or second-order logic is the proposition that every set *A* determines a full structure, i.e. one that contains every relation in extension of every arity on *A*; such a structure forms the basis for a standard model for the language of second- or higher-order logic.)

It is important to be clear on the nature of the difficulty Newman has uncovered. The problem is *not* a failure of the theory to specify the domain of objects on which a model of our theories of the world is to be defined. The difficulty is not the one of Pythagoreanism raised by Quine (1964) and Winnie (1967, Theorem 2, pp. 227–9); that is to say, it is not that Russell cannot rule out abstract models. Indeed Russell himself raises this problem in the introduction to *The Analysis of Matter*:

It frequently happens that we have a deductive mathematical system, starting from hypotheses concerning undefined objects, and that we have reason to believe that there are objects fulfilling these hypotheses, although, initially, we are unable to point out any such objects with certainty. Usually, in such cases, although many different sets of objects are abstractly available as fulfilling the hypotheses, there is one such set which is much more important than the others . . . The substitution of such a set for the undefined objects is "interpretation." This process is essential in discovering the philosophical import of physics.

The difference between an important and an unimportant interpretation may be made clear by the case of geometry. Any geometry, Euclidean or non-Euclidean, in which every point has co-ordinates which are real numbers, can be interpreted as applying to a system of sets of real numbers – i.e. a point can be taken to *be* the series of its co-ordinates. This interpretation is legitimate, and is convenient when we are studying geometry as a branch of pure mathematics. But it is not the important interpretation. Geometry is important, unlike arithmetic and analysis, because it can be interpreted so as to be part of applied mathematics – in fact, so as to be part of physics. It is this interpretation which is the really interesting one, and we cannot therefore rest content with the interpretation which makes geometry part of the study of real numbers, and so, ultimately, part of the study of finite integers. (*AM*, pp. 4–5)

In this case we have a simple criterion for separating important from unimportant interpretations: the important interpretations are connected with our observations, the unimportant ones are *not*. Newman's problem arises *after* the domain has been fixed: in the case of abstract versus physical geometry we have to distinguish different relations on different domains.

The problem is made easier by the fact that we can exclude one of the domains *and therefore* one class of relations. When the domain of the model is *given*, we must "distinguish between systems of relations that hold among the members of a given aggregate" (Newman 1928, p. 147). This is a difficulty because there is *always* a relation with the structure W. Russell cannot avoid trivialization by claiming that the relation with structure W that exists as a matter of logic is not necessarily the *important* relation with structure W (as the interpretation in $\mathfrak{R}^3$ is not the important interpretation of geometry). That is to say, one cannot avoid trivialization in this way without some means of distinguishing important from unimportant relations on a given domain. But the notion of importance that must be appealed to is one for which Russell can give no explanation, and in fact his own theory precludes giving any explanation of the notion. Newman's summary is well worth quoting in full:

In the present case [i.e. in the case where we must choose among relations on a given domain rather than choose among different relations on different domains] we should have to compare the importance of relations of which nothing is known save their incidence (the same for all of them) in a certain aggregate. For this comparison there is no possible criterion, so that "importance" would have to be reckoned among the prime unanalyzable qualities of the constituents of the world, which is, I think, absurd. The statement that there is an *important* relation which sets up the structure W among the unperceived events of the world cannot, then, be accepted as a true interpretation of our beliefs about these events, and it seems necessary to give up the "structure–quality" division of knowledge in its strict form. (Newman 1928, p. 147)

Recall that giving up the structure–quality division in its strict form means giving up the idea that we do not know the qualities of unperceived objects. This of course would rob Russell's theory of its distinctive character, and so far as we can tell, Russell remained equivocal on this issue for the remainder of his philosophical career.

It might be thought that Newman's objection depends on too extensional an interpretation of relations, and that once this is given up, a way out of the difficulty is available. The idea is that logic (or logic plus set theory) only guarantees what Newman calls *fictitious* relations – relations whose only extrinsic property is that they hold between specified pairs of individuals – e.g. binary relations whose associated propositional functions are of the form *x is a and y is b, or x is a and y is c, etc.*, where *a*, *b*, and *c* are elements of *A*. But in fact this is not so. Given some means of identifying the individuals of the domain *A*, we can always find an isomorphic structure

W* which holds of a nonfictitious relation R^* on a domain A^*. We may then define a nonfictitious relation S whose field is included in A: S is just the image of R^* under the inverse of the isomorphism that correlates A and A^*.

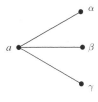

Figure 5.1

Newman gives a simple combinatorial example. Consider the graph $\Gamma(R)$ of a relation R on $A \times A$, $A = \{a, \alpha, \beta, \gamma\}$, depicted in Figure 5.1. Define the new relation S on A by letting S hold of (x, y) if $f(x)$ and $f(y)$ belong to different alphabets where 'a' $= f(a)$, 'α' $= f(\alpha)$, 'β' $= f(\beta)$, 'γ' $= f(\gamma)$. Then S is nonfictitious. It is clear that this strategy is perfectly general. Following Newman, call a nonfictitious relation *real*. Then reality in this sense is "preserved under isomorphism"; i.e. any extensionally specified relation may be regarded as the image of a real (nonfictitious) relation. And this real relation exists if the fictitious one does. Thus a claim of the form 'There exists a *real* relation R such that . . .' will be true given a claim of the form 'There exists a relation R such that . . .' Of course, S may nonetheless be *trivial* or *unimportant*. But this just shows that eliminating extensionalism does not solve our problem.*

We might try to add further constraints on the notion *R is a real relation*. But then we move away from the relatively clear intuition that motivated its original characterization: namely, that a relation with only a purely extensional characterization is fictitious. (Notice that certain obvious candidates such as projectability cannot be appealed to since it is precisely the class of relations and properties that enter into laws that we are attempting to delimit.)

---

* *Note added in 2012*: An elaboration of the response to Newman considered in the text has recently gained currency. It holds that the scope of the quantified variable should be restricted to *natural* or *real* relations in a stronger sense than Newman's rejoinder assumes. It is by no means obvious that a condition . . . *R* . . . is satisfied by a *real* or *natural R* in this sense. However, whatever the notion of naturalness or reality, there are two difficulties with such a response. First, it makes a simple fact involving our judgments of epistemic significance depend on metaphysically difficult notions when it is evident that it does not. And secondly, it fails to capture the scope of Newman's observation, which is independent of a relation's "reality" or "naturalness," since there is a distinction to be made between significant and nonsignificant claims regarding the structure of a relation even when the relation is wholly artificial.

The conclusion Newman draws from this analysis is, we think, the right one: since it is indisputably true that our knowledge of structure is nontrivial – we clearly do not stipulate the holding of the structural properties our theories postulate – it cannot be the case that our knowledge of the unperceived parts of the world is *purely* structural. The unobservable–observable dichotomy is not explicable in terms of the structure–quality division of knowledge without giving up the idea that our knowledge of the unobservable parts of the world is discovered rather than stipulated. Of course it is also possible to give up the naive intuition on which Newman's analysis depends. In this case one would *accept* the stipulative character of the theoretical components of our knowledge, and indeed much of the philosophy of science that immediately preceded and subsequently followed *The Analysis of Matter* did just this. Russell is unique in wanting to preserve a nonconventionalist view of the world's structure, while retaining a structure–quality or form–content division of knowledge that is intended to more or less correspond to the division between theoretical and observational knowledge.

By the way, Newman's objection also explains the intuition that, despite its intention, Russell's structuralism collapses into phenomenalism. The difficulty is that every assertion about unperceived events is trivially true, i.e. true as a matter of logic plus an empirical assumption concerning cardinality. Now, the phenomenalist claims that statements about the external world are "reducible" to statements about perception. Whatever *else* this may mean, it requires that, given the appropriate reductive definitions, statements about the physical world are entailed by statements about perception. But if Russell's view is accepted, this characteristic consequence of phenomenalism is almost guaranteed: if statements about the physical world are true as a matter of logic, then they are implied by every proposition, and, in particular, they are implied by the statements of perception. Thus modulo a single nonlogical assumption concerning the cardinality of the external world, Russell's structuralism guarantees the truth of this phenomenalist thesis.

The only reference to Newman's paper of which we are aware is in the second volume of Russell's autobiography. Russell very graciously acknowledges Newman's criticism in a letter he reprints without comment. The letter is worth quoting at length. Dated April 24, 1928, it is identified simply as "To Max Newman, the distinguished mathematician."

Dear Newman,
Many thanks for sending me the off-print of your article about me in *Mind*. I read it with great interest and some dismay. You make it entirely obvious that my statements to the effect that nothing is known about the physical world except its

structure are either false or trivial, and I am somewhat ashamed at not having noticed the point for myself.

It is of course obvious, as you point out, that the only effective assertion about the physical world involved in saying that it is susceptible to such and such a structure is an assertion about its cardinal number. (This by the way is not quite so trivial an assertion as it would seem to be, if, as is not improbable, the cardinal number involved is finite. This, however, is not a point upon which I wish to lay stress.) It was quite clear to me, as I read your article, that I had not really intended to say what in fact I did say, that *nothing* is known about the physical world except its structure. I had always assumed spatio-temporal continuity with the world of percepts, that is to say, I had assumed that there might be co-punctuality between percepts and non-percepts, and even that one could pass by a finite number of steps from one event to another compresent with it, from one end of the universe to the other. And copunctuality I regarded as a relation which might exist among percepts and is itself perceptible.

(Russell 1968, p. 176)

To our knowledge, Russell never discusses the puzzle in any of his later work. He seems to give up the idea that our knowledge of the physical world is purely structural, but there is no account of how, on his theory of knowledge (e.g. the theory developed in *Human Knowledge: Its Scope and Limits*), such nonstructural knowledge can arise. Yet all the elements of the earlier and later theories are the same – the only difference is in the conclusion drawn. Thus either the original claim (that we are restricted to purely structural knowledge) was theoretically unmotivated or the argument of the later theory contains a lacuna. The difficulty in adjudicating between these alternatives is that the theoretical development is made to depend on what we regard as falling within "acquaintance." And this makes the resolution quite artificial: in the earlier theory we could not assume acquaintance with what Grover Maxwell called a "cross category notion" such as spatio-temporal contiguity or causality, but in light of the difficulties of that theory we now find that we *can* assume this.[13] We are not saying that one cannot resolve the issue in this way. But it seems quite clear that without a considerable advance in the theoretical articulation of this rather elusive Russellian concept, no such resolution of the difficulty can be very compelling.

5.3

There is another consequence that Newman draws, although it is much less explicit in the article. Russell's structuralism can be viewed as a theory of how the reference of the theoretical vocabulary is fixed. On this view the

---

[13] Although Maxwell opted for this solution in the papers cited earlier, unlike Russell, he was explicit about the nature of the problem.

reference of a T-predicate is fixed by its connection with observation, and its structural properties. As we have seen, the connection with observation fixes the domain (or at least establishes that very many sets are *not* the domain of interpretation). Structure is supposed to complete the task while preserving the nonstipulative character of the truth of our theoretical knowledge. As we saw, Newman observes that within Russell's theory, there is no analogue for relations of the important–unimportant criterion for domains. From a contemporary, model-theoretic standpoint, this is just the problem of intended versus unintended interpretations: Newman shows that there is always some relation R (on the intended domain) with structure W. But if the only constraints on something's being the intended referent of 'R' are observational and structural constraints, no such criterion for distinguishing the intended referent of 'R' can be given; so that the notion of an intended interpretation must, in Quine's phrase, be provided by our background theory, and hence cannot be a formal or structural notion in Russell's sense. Now, in fact, something strikingly similar to this argument has recently been rediscovered by Hilary Putnam, and it has been used by him to pose a puzzle for certain "realist" views of reference and truth.

Putnam's formulation of the argument is model-theoretic and general rather than informal and illustrative. Suppose we are given a theory T, all of whose observational consequences are true. We assume, for the sake of argument, that the observational consequences of T may be characterized as a subset of the sentences generated from a given "O-vocabulary." It is assumed that the interpretation of the language L(T) of T is specified for its O-vocabulary, so that T is "partially interpreted" in the sense that the interpretation function is a partial function. Assume that T is consistent and that M is an abstract model of the uninterpreted sentences of T of the same cardinality as T's intended domain. Following Putnam we let 'THE WORLD' denote the intended domain of T. Now extend the interpretation to the theoretical vocabulary of T by letting each predicate of the theoretical vocabulary of T denote the image in THE WORLD of its interpretation in M under any one-to-one correspondence between M and THE WORLD. Call this interpretation (which extends the partial interpretation) 'SAT.' Clearly SAT is completely arbitrary, and should be an utterly unacceptable extension since it trivializes the question whether T is true. Any theory of knowledge and reference that is incapable of rejecting SAT, i.e. any theory that is incapable of distinguishing truth from truth under SAT – TRUTH(SAT) – is committed to the implication, T is true if T is TRUE(SAT). But T is TRUE(SAT) as a matter of logic. That is, T is TRUE(SAT) follows, via the completeness theorem, from 'T is consistent.'

So any theory that cannot exclude SAT as an intended interpretation of L(T) cannot account for our naive confidence in the belief that our theories, if true, are "significantly true."[14]

Except for its metalogical flavor, this argument parallels Newman's. Where Putnam argues from the consistency of a set of sentences, Newman argues directly from the existence of a relational structure satisfying the intuitive conditions of a model. (Recall that Newman was writing in 1928 and in the logicist tradition of Whitehead and Russell – a tradition to which the metalogical turn in mathematical logic was quite alien.)[15] The chief difference between Putnam and Newman is in the use made of the argument. Putnam, as we have said, employs the argument against a certain form of realism – one which is rather difficult to isolate, although there are some suggestions. For example, Putnam says it is that form of realism which holds that even an epistemically ideal theory might be false; or any view which regards truth as radically nonepistemic; or any theory of truth and reference which "treats language as a 'mirror of the world,' rather than one which supposes merely that 'speakers mirror the world' – i.e. their environment – in the sense of constructing a symbolic representation of that environment."[16] All these remarks, though certainly suggestive, are elusive.

In *Reason, Truth and History*,[17] Putnam expresses the significance of the result slightly differently: he characterizes the argument as supporting the contention of Quine (1969) that fixing the set of true observation sentences together with the set of accepted theoretical sentences will not determine the reference of the theoretical vocabulary. Putnam argues that the indeterminacy affects the reference of the T-vocabulary in every possible world, and in this respect extends Quine's claim.[18] The formal connection with Newman's argument is this. Newman shows that fixing the domain and models up to isomorphism does not fix the intended reference of the T-vocabulary. Putnam shows that fixing the domain and models up to elementary equivalence does not fix the intended reference of the T-vocabulary. (Thus formally Newman's conclusion is stronger than

[14] John Winnie (1967) employs an argument of this form in the proof of his Theorem 1. Winnie's discussion focuses on formal features of the puzzle: that there is not a unique interpretation of the theory and that there is always an arithmetical interpretation. Newman and (we believe) Putnam emphasize a different point: namely, that *mere consistency* seems to be sufficient for *truth*; i.e. since T has an arithmetical interpretation it has a true interpretation *in the world* (i.e. in THE WORLD).
[15] See Goldfarb (1979) for a discussion.
[16] Putnam (1978, p. 123). The other suggestions may also be found in this chapter, which is a reprinting of Putnam (1977) – Putnam's presidential address to the American Philosophical Association.
[17] Putnam (1981, pp. 33–4).   [18] See the appendix to Putnam (1981).

Putnam's, since isomorphism strictly implies equivalence, although we do not wish to lay much stress on this fact.)

It may be that Newman's argument contains an important observation about realism *in general* – not merely about *structural* realism. The recent work of Putnam, to which we have drawn attention, is intended to suggest that something very much like it does. But we have been unable to find or construct a clear statement of the connection. Certainly the argument poses difficulties for partial interpretation accounts of theories that exactly parallel the difficulties confronted by Russell's theory: both accounts of theoretical knowledge and reference fail to square with our naive beliefs regarding the nature of the truth of theoretical claims. The conclusion we have endorsed is a conservative one: the argument shows that neither Russell nor the neo-positivist doctrine of partial interpretation has gotten the analysis of theoretical knowledge and reference quite right. A satisfactory account of these notions must do justice to such "obvious facts" as that the world's structure is discovered rather than stipulated. We are skeptical of our ability to do this on any view of reference and truth that does not take the language of physical theory as the "ultimate parameter" within which reference is fixed.

5.4

The difficulty Newman raises for Russell's use of the notion of structure can be clarified and deepened if we look at an analogous difficulty that arises for Carnap in the *Aufbau* and then take a brief glance at Wittgenstein's *Tractatus*. Newman's problem can be put this way. Russell wishes to turn theoretical terms into variables by ramsification; accordingly, physics becomes the assertion that there *exist* properties and relations having certain logical features, satisfying certain implicit definitions. The problem is that this procedure trivializes physics in the sense that it threatens to turn the *empirical* claims of science into mere *mathematical* truths. More precisely, if our theory is consistent, and if all its purely observational consequences are true, then the truth of the Ramsey sentence *follows* as a theorem of set theory or second-order logic, provided our initial domain has the right cardinality – if it does not, then the consistency of our theory again implies the existence of a domain that does.[†] Hence, the truth of physical theory reduces to the truth of its observational consequences, and as we saw earlier, Russell's realism

---

[†] *Note added in 2012*: This formulation overlooks a subtlety that is discussed in Sections 7.2 and 7.4.

collapses into a version of phenomenalism or strict empiricism after all: *all* theories with the same observational consequences will be equally true.

We have observed that Carnap attempts to go one better than Russell in the *Aufbau* by turning *all* scientific terms into variables by means of purely structural definite descriptions. So Carnap's program is faced with a closely analogous difficulty, a difficulty he himself articulates with characteristic honesty and precision in sects. 153–5. Up to this point in the *Aufbau*, all scientific concepts have been introduced via structural definite descriptions containing the single nonlogical relation R_s of recollected similarity, and the visual field, for example, is the unique five-dimensional "sense class" *based on R_s*. However, the goal of "complete formalization" – the goal of converting all definitions into *purely* structural definite descriptions – can only be achieved if the basic relation R_s is itself eliminated (sect. 153). How is this to be done? Well, R_s also has a logical form or structure: it has a graph $\Gamma(R_s)$. For example, R_s is asymmetrical (sect. 108), and there is one and only one "sense class" based on R_s that is five-dimensional (sects. 117–19). To be sure, we can know these formal properties of R_s only empirically, but once we know them, we can express them in a purely formal schema $\Gamma(R)$, where R is now a relation *variable*. The idea, then, is to define R_s as the unique relation with this graph:

$$R_s = (\imath R)\Gamma(R). \tag{3}$$

But as Carnap immediately points out, there is a serious problem with this procedure. Since $\Gamma(R)$ is now a purely formal schema containing no nonlogical primitives, the *uniqueness* claim implicit in (3) will never be satisfied. As long as the condition $\Gamma(\ldots)$ is consistent, there will always be infinitely many distinct relations which satisfy it, even if we confine ourselves to relations on the field of R_s:

Our assumption is justified only if the new relation extensions are not arbitrary, unconnected pair lists, but if we require of them that they correspond to some experienceable, "natural" relations (to give a preliminary, vague expression).

If no such requirement is made, then there are certainly other relation extensions [besides R_s itself] for which all constructional formulas can be produced ... All we have to do is carry out a one-to-one transformation of the set of basic elements into itself and determine as the new basic relations those relation extensions whose inventory is the transformed inventory of the original basic relations. In this case, the new relation extensions have the same structure as the original ones (they are isomorphic). (sect. 154)

The analogy with the Newman problem is clear. How does Carnap handle the problem? He introduces the notion of *foundedness* to characterize those relation extensions that are "experienceable and 'natural,'" and argues that one can view this notion as a primitive concept of logic:

[Foundedness] does not belong to any definite extralogical object domain, as all other nonlogical objects do. Our considerations concerning the characterization of the basic relations of a constructional system as founded relation extensions of a certain kind hold for every constructional system of every domain whatever. It is perhaps permissible, because of this generality, to envisage the concept of foundedness as a concept of logic and to introduce it, since it is undefinable, as a *basic concept of logic* . . . [L]et us introduce the class of founded relation extensions as a basic concept of logic (logistic symbol: *found*) without considering the problem as already solved. (sect. 154)

In other words, Carnap proposes to do precisely what we saw Newman recoil from: he wants to reckon "importance" (foundedness) as "among the prime unanalyzable qualities of the constituents of the world"! Once this is done, the problem of defining R_s is easily solved. We replace (3) with

$$R_s = (\iota R)(found(R) \,\&\, \Gamma(R)) \qquad (4)$$

Actually, it is not entirely obvious why Carnap needs a definition like (3), because it is not clear why he needs to make a uniqueness claim. He could express our empirical knowledge, as Russell does, with a purely existential claim

$$\exists R \Gamma(R)) \qquad (5)$$

where (5) fulfills the goal of "complete formalization" just as well as (3), in so far as both contain no nonlogical vocabulary. Of course (5) is vulnerable to the Newman objection in precisely its original form – it again threatens to turn empirical truth into mathematical truth – but let us slow down and reflect for a moment. For, when applied to our current physical theories, the assumptions needed to prove (5) turn out to be rather strong. In particular, we will need some version of the Axiom of Infinity and some version of the power set axiom applied to infinite sets (that is, we will need the continuum). Contemporary model theory for second-order logic makes both these assumptions, and that is why (5) is a *theorem*. Suppose, however, that we are transported back to the 1920s. The line between logic and mathematics is much less clear, and it is equally unclear whether strong existence assumptions like the Axiom of Infinity and the power set axiom are to be counted as part of logic *or* mathematics.[19] (In *Principia Mathematica*, for example, Russell carries along the Axiom of Infinity as an undischarged hypothesis.) Accordingly, the status of (5) is much less clear-cut.

To see some of the issues here, consider the strictest and most rigorous version of logicism devised in this period: namely, Wittgenstein's *Tractatus*. The *Tractatus* is clear and emphatic about the Axiom of Infinity and set theory: neither is part of logic (5.535, 6.031). In fact, logic makes no cardinality claims whatsoever since there are no privileged numbers (5.453). By the same token, however, such cardinality claims are not part of mathematics either, for mathematics is a "logical method" involving only the simplest forms of combinatorial equations (6.2–6.241). Although Wittgenstein is not very explicit in the *Tractatus* about the precise scope of mathematics, it is apparently exhausted by a rather small fragment of elementary arithmetic. On this kind of conception, then, formulas like (5) will *not* be theorems of logic or mathematics; if true at all, they can only be empirically (synthetically) true. Hence, on a Tractarian conception of logic and mathematics, we could perhaps make sense of Carnap's program of "complete formalization" after all. And for all we know, Wittgenstein could have had something very much like this in mind. (See *Tractatus* 5.526– 5.5262, for example, where Wittgenstein says that we can describe the world completely by means *of fully generalized* propositions. See also 6.343– 6.3432.) In any case, whether or not these considerations have any connection with Wittgenstein's intentions in the *Tractatus*, they do illustrate some of the intellectual tensions produced by logicism's attempt to account simultaneously for both pure mathematics and applied mathematics (mathematical physics). In general, it appears that we can account for the distinctive character of one only at the expense of the other. This is perhaps the final lesson of the Newman problem.[‡]

---

[19] The *Aufbau* is remarkably sanguine about this. Carnap is apparently willing to count both *AxInf* and *PowerSet* as logical. In sect. 125, for example, he introduces an *n*-dimensional real-number space with a completely innocent air. Apparently, then, he is willing to count virtually all of set theory as logic (see also sect. 107). This is why (5) could not possibly serve as an expression of our *empirical* knowledge for Carnap.

[‡] *Note added in 2012*: I no longer consider this a plausible conjecture about logicism; indeed, the four previous chapters are an extended argument against it. But I think something close to it is true of Carnap's Ramsey-sentence reconstruction of the language of science. Nor do I agree with the statement, made earlier in the paragraph, that the *Tractatus* is the most rigorous version of logicism devised in this period.

CHAPTER 6

On the rational reconstruction of our theoretical knowledge

6.1 INTRODUCTION

My focus in this chapter is the rational reconstruction of physical theories initially advanced by F. P. Ramsey, and later elaborated by Rudolf Carnap. As will become clear in the sequel, the Carnap–Ramsey reconstruction of theoretical knowledge is a natural development of classical empiricist ideas, one that is informed by Russell's philosophical logic and his theories of propositional understanding and knowledge of matter; as such, it is not merely a schematic representation of the general notion of an empirical theory, but the backbone of a general account of our knowledge of the physical world. Nor is it merely an interesting episode in the history of the philosophy of science: Carnap–Ramsey is an illuminating, if not ultimately satisfying, approach to epistemological problems that remain with us.

To give a preliminary overview, the classical epistemological issue to which Russell sought to apply his logical discoveries was that of showing how, on the basis of minimal assumptions regarding the scope of our experience, we might articulate an account of our knowledge of the material world. A first step toward a solution would have to address the fact that we succeed in *understanding* propositions which transcend the limitations of our experience. It is how Russell addressed this first step that is of primary interest for what follows. In the reconstructive program of Carnap and Ramsey, Russell's epistemological problem is transformed into an issue in the theory of theories: on the basis of a particular choice of a minimal nonlogical vocabulary, to recover our theoretical knowledge within an expressively equivalent framework, a framework that preserves the characteristic features of our applications of the concepts of logical consequence, reference, and truth. It is a condition of adequacy accepted by both programs that they should recover many pre-analytic intuitions about our

First published in *British Journal for the Philosophy of Science* 54 (2003): 371–403.

theoretical knowledge. Thus, although both the classical and the reconstructive programs have foundationalist overtones, it would be a mistake to view
them as motivated by skeptical doubts concerning our theoretical beliefs.

In light of the extreme generality of the Carnap–Ramsey reconstruction,
it might seem tendentious to characterize it as a reconstruction of *physics*.
Indeed, the highly schematic and abstract style that the approach exempli-
fies has given way to more specialized foundational investigations of particular classes of physical theories. Unfortunately, this shift was accomplished
without sufficient appreciation of what the reconstructive program we will
be examining sought to accomplish. The aims of the Carnap–Ramsey
reconstruction mandated that it should be stated with great generality;
even though the reconstruction does not depend on the characteristics of
any special class of physical theories, its applicability to physics is essential to
its epistemological point. I believe – and will try to show – that insuperable
difficulties confront the Carnap–Ramsey reconstructive program. But I
hope also to make it clear that Carnap's and Ramsey's reconstructions
possess not only internal coherence and elegance, but, more importantly,
a degree of philosophical motivation not matched by rival accounts. Let me
begin with a review of the relevant Russellian background to the program I
will be exploring.

6.2 RUSSELL'S THEORY OF PROPOSITIONAL UNDERSTANDING

Russell's first reasonably well-articulated application of his theory of
descriptions to a traditional epistemological problem – namely that of
determining the nature and scope of our knowledge of matter – occurs in
The Problems of Philosophy[1] where the theory is extended by the addition of
the description theory of names and deployed in support of an exceptionally
simple theory of propositional understanding or theory of meaning. Putting
to one side the issue of vacuous names, Russell's theory of meaning tells us
that if a sentence $S(n)$ contains a name n for an individual with whom we are
not acquainted, the proposition expressed by the sentence cannot contain
the bearer of the name among its constituents. We must instead imagine
that the name is shorthand for a description. This description is in turn
analyzed – "contextually defined" after the fashion of the theory of descriptions – into expressions for individuals and propositional functions which
are proper constituents of the proposition expressed. The individuals and

---

[1] Russell (1912); cited as *Problems*.

propositional functions are so chosen that the resulting proposition satisfies what in *Problems* Russell called "the fundamental principle in the analysis of propositions containing descriptions: *Every proposition which we can understand must be composed wholly of constituents with which we are acquainted*" (*Problems*, p. 58).

Notice that the point of the theory of *Problems* is not to eliminate what a nonvacuous name stands for, but to explain, compatibly with the fundamental principle, how a sentence containing the name is understood.[2] For example, I understand the sentence, 'Bismarck was an astute diplomat' because I am acquainted with a propositional function that only Bismarck satisfies. A useful way of putting the matter is to say that although the proposition I *assert* with this sentence contains Bismarck as a constituent, and I succeed in saying something *about* him, he is not a constituent of the proposition I *express*.[3] By hypothesis, the bearer of the name 'Bismarck' exists and is the unique individual satisfying some (possibly complex) propositional function-expression. However, not being known by acquaintance, Bismarck is not *himself* a constituent of the proposition expressed. He is nevertheless someone to whom we are able to refer and make assertions *about* because of our acquaintance with a property only he has. If, for

---

[2] The exposition of this point is made difficult by the fact that Russell is not always consistent about his use of 'contextual definition.' In fact, Russell can be quite *un*clear on the distinction between the contextual analysis of an incomplete symbol and the explicit definition of an entity or "complete symbol" – occasionally even equating the two notions. Cf., for example, the following passage from *Logical Atomism*:

> One very important heuristic maxim which Dr Whitehead and I found, by experience, to be applicable in mathematical logic, and have since applied in various other fields, is a form of Ockham's razor ... The principle may be stated in the form: 'Wherever possible, substitute constructions out of known entities for inferences to unknown entities' ... A very important example of the principle is Frege's definition of the cardinal number of a given set of terms as the class of all sets that are 'similar' to the given set ... Thus a cardinal number is the class of all those classes which are similar to a given class. This definition leaves unchanged the truth-values of all propositions in which cardinal numbers occur, and avoids the inference to a set of entities called 'cardinal numbers,' which were never needed except for the purpose of making arithmetic intelligible, and are now no longer needed for that purpose ... Another important example concerns what I call 'definite descriptions,' i.e., such phrases as 'the even prime' 'the present King of England,' 'the present King of France.' There has always been a difficulty in interpreting such propositions as 'the present King of France does not exist.' The difficulty arose through supposing that 'the present King of France' is the subject of this proposition ... The fact is that, when the words 'the so-and-so' occur in a proposition, there is no corresponding single constituent of the proposition, and when the proposition is fully analyzed the words 'the so-and-so' have disappeared. (Russell 1924, pp. 326–8)

Russell has here overlooked the fact that there must be some independent motivation for treating something as an incomplete symbol – something more than the mere possibility of applying the method of contextual definition.

[3] For a fuller elaboration of the distinction between asserting a proposition and expressing it, and the role this distinction plays in Russell's theories of propositional understanding, see Demopoulos (1999).

example, Bismarck is identified as the first Chancellor of the German Empire, then I succeed in making assertions about him because, among other things, I am acquainted with the relation expressed by '*x* is Chancellor of *y*.'

Russell's argument against Berkelian idealism – clearly one of the central lessons of *Problems* – is based on the observation we have just reviewed regarding his description theory of names. On Russell's reconstruction, Berkeley fallaciously assumed that the fundamental principle restricts what we can have knowledge about – what propositions we can *assert* – when, in fact, it restricts only the propositions we can *express*.

The application of Russell's new theories of propositions and denoting to our knowledge of the material world proceeds from three explicit assumptions and one tacit assumption. The explicit assumptions are: (i) we are not acquainted with matter; but (ii) it is always possible to formulate a description which is uniquely satisfied by the material object to which we take ourselves to refer; and (iii) these descriptions involve only propositional functions and individuals with which we are acquainted. The tacit assumption is that (iv) the propositional functions with which we are acquainted can be so chosen that their logically primitive constituents apply only to individuals with which we are acquainted in perception. Without the tacit assumption (iv), the critique of Berkeley and the epistemological significance of Russell's theory would be severely limited, since it could be objected that the view secures realism about one part of the material world only relative to realism about another.

Precisely how the tacit assumption is to be satisfied is something *Problems* only hints at – for example, when Russell writes:

[I]f a regiment of men are marching along a road, the *shape* of the regiment will look different from different points of view, but the men will appear arranged in the same order from all points of view. Hence we regard the *order* as true also in physical space, whereas the shape is only supposed to correspond to the physical space so far as is required for the preservation of the order. (*Problems*, pp. 32–3)

Russell's full elaboration of this idea is given in *The Analysis of Matter*,[4] where his "structuralism" is articulated at length. What the response to Berkeley requires is a theory that will allow us to dispense with primitive nonlogical vocabulary items which name anything with which we are not perceptually acquainted while allowing that we can have knowledge *about* things which fall outside the realm of our perceptual acquaintance. In particular, in its application to our knowledge of matter, we demand a

---

[4] Russell (1927).

theory that will explain how our ability to formulate propositions which express truths about the material world need not in any way require our acquaintance with that world. It is these desiderata that Russell's structuralism was intended to fulfill.

To see at least in outline how structuralism proposed to meet these goals, let us recall what is characteristic of the general characterization of structure in terms of structural similarity and its elaboration in the "relation arithmetic" of *Principia Mathematica*.[5] The model on which Russell's definition of a structure was based is the Frege–Russell definition of the cardinal numbers as similarity classes under the relation of one–one correspondence. As Frege perceived in *Grundlagen*,[6] and as Russell was to discover some years later, the notion of one–one correspondence, being definable in wholly logical terms, is independent of spatio-temporal intuition. It follows that this must also be true of *structural similarity*, since it rests only on the notion of one–one correspondence and the general concept of a relation. Thus, for Russell, the *philosophical* interest of structural similarity derives from the fact that, as a notion of pure logic, it owes nothing to experience or Kantian intuition.

From very early on Russell seems to have seen his account of structure as capable of providing a framework within which it would be possible to articulate the nature of the similarity philosophers had supposed to exist between appearance and reality or, to use Kantian terminology, between the phenomenal and the noumenal worlds. The significance of this point was not lost on Russell, nor, I dare say, was the irony that a concept which owed its genesis to logicism might usefully contribute to the articulation of a Kantian doctrine. What had defeated previous attempts was the want of a notion of similarity which was not so great that it would collapse the gulf that was supposed to exist between them, and was not so slight that it could not be reckoned a significant sense of similarity. Russell believed that with the discovery of the notion of structural similarity, he had solved this metaphysical and epistemological problem.

Russell's picture of how the application to Kant should go appears to have been something like this. The noumenal world, not being given to us in intuition, cannot, apparently, be required to have properties in common with the phenomenal world. This leaves us with the problem of understanding how to formulate any conception of what the noumenal world is like and of understanding how it can be knowable. But because structural similarity has a purely logical characterization, it is independent of

---

[5] Whitehead and Russell (1912).   [6] Frege (1884).

intuition. The noumenal world thus emerges as an isomorphic copy of the phenomenal world, one which we may suppose has the requisite similarity with the world of phenomena without thereby committing ourselves to the idea that it shares any of the *intuitive* properties of the phenomenal world. Had it not proved possible to capture this notion of similarity by purely logical means, we would have been precluded from assuming even this degree of similarity between noumena and phenomena, and might, therefore, have been inclined toward some form of idealism regarding the world behind phenomena. The purely logical notion of structural similarity preserves us from this tendency toward idealism, since it shows how we might have knowledge about the relations that order the noumenal world, without needing to assume intuitive knowledge of those relations.[7]

It is clear that the same thought underlies our earlier quote from *Problems* regarding *shape* and *order*: it is not necessary to suppose *acquaintance* with the relation that holds among the members of the regiment provided we can express the *structure* of that relation, as, indeed, the notion of order permits us to do. Similarly, we need to be able to express the structure of the relations that hold among events in the material world, and on this basis, one might hope to construct the whole "spatio-temporal" framework appropriate to the world behind phenomena. By means of this framework we can proceed to describe all those material events with which we are not acquainted, but regarding which we wish to *assert* many propositions. But to achieve this we need not be acquainted with the relations that generate this framework.

This is an elegant application of a technical idea of mathematical logic to a philosophical problem. It is, however, subject to an important limitation. If we intend the statement that the noumenal world is isomorphic to the phenomenal one to be more than part of its *definition* – if, that is, we also intend it to be a significant claim that the noumenal and phenomenal worlds are structurally similar – then we are implicitly assuming that we have access to the relations holding among things in themselves independently of the correspondence in terms of which their similarity to phenomenal relations has been characterized. Otherwise, the observation that structural similarity allows us to preserve the comparability of the noumenal and phenomenal worlds is based on a mere stipulation, the character of the noumenal world having been *defined* in terms of the isomorphism. Since it

---

[7] As noted in the text, structuralism is developed at length in *The Analysis of Matter* (Russell 1927), but this application of the view was already announced in *Introduction to Mathematical Philosophy* (1919, pp. 61–2) quoted earlier in Section 5.1.

owes nothing to intuition, structural similarity may be used to address the objection that there is *literally* nothing that can be said regarding things in themselves. But the conception of the noumenal world to which we are led has the unintended consequence that its similarity to the phenomenal world is the result of a definition.

The nature of a claim of structural similarity is such that it is a significant assertion only when the relations being compared are given independently of the mapping which establishes their similarity. For this we require more than an appropriately general notion of similarity; we must, in addition, have independent knowledge of the relations between which the similarity is supposed to hold. Knowledge of the relations among things in themselves cannot be purely structural, they cannot merely be known as the images, under a suitable mapping, of the relations among phenomena, since that would make the claim of their similarity to relations among phenomena a tautology. But, things in themselves being "in themselves," neither can it be intuitive. The noumenal world would seem, therefore, to have retained almost all of its elusive character. Mutatis mutandis, the relations among events of the physical world cannot be defined as the isomorphic images of the relations holding among those events with which we are acquainted if we intend to preserve the nontriviality of the thesis that the two systems of relations are similar. As we will see, this difficulty is pervasive, and is one that reemerges in a sharper form in the context of later developments.

Finally, let us note by way of conclusion that Russell's use of the distinction between acquaintance and description rests on the assumption that although knowledge by acquaintance is not required for us to have knowledge about some entity or other, knowledge of a propositional function uniquely satisfied by that entity suffices to ensure *reference* to it. This assumption is questionable if we think of knowledge of reference as something that carries with it knowledge of *what* (or of *whom*, etc.) we are referring to. In fact, Russell does *not* think that such knowledge is required for reference to be successful. Instead, he endorses the view that having such knowledge – 'knowledge-wh,' as I shall call it – requires knowledge by acquaintance. But acquaintance seems to be far too strong a requirement for knowledge-wh, if the pre-analytic connection between knowledge-wh and knowledge of reference is to be preserved. It would thus seem that knowledge-wh is a type of knowledge that is not captured by the acquaintance–description division: I know that the next president of the United States will be the person who gets the most Electoral College votes, and I know that *x gets the most Electoral College votes in the US election of 2000* is a propositional function that, when satisfied, is uniquely satisfied, but it is at least questionable whether, knowing these things and using this expression, I succeed in *referring* to anyone.[8] And while I am not acquainted with George W. Bush, it is plausible to suppose that I know who he is, and thus satisfy one condition for sometimes successfully referring to him. In short, knowledge by description appears to be too "thin," and knowledge by acquaintance too "thick," to tell us what knowledge of reference consists in. Perhaps Russell thought that further restricting the propositional constituents with which we refer to things outside our acquaintance to constituents with which we are perceptually acquainted overcomes – rather than compounds – the difficulty with this use of knowledge by description. If so, he was mistaken, and the extent of his mistake will become apparent from our discussion of the extension of his ideas to the theory of theories.

6.3 RAMSEY'S PRIMARY AND SECONDARY SYSTEMS

I have been careful to present Russell's elaboration of his theory of propositional understanding in such a way that its connection with a subsequent development by Ramsey will be transparent. A feature of Russell's theory that I have alluded to is the technique by which it avoids the use of a name for something which is not an object of acquaintance. It is this consequence of Russell's deployment of his theory of descriptions that was imitated by Ramsey, in his posthumously published "Theories."[9] In this paper Ramsey proposed that the "content" of a physical theory can be captured by what has come to be known as its Ramsey sentence. It will be recalled that the Ramsey sentence $R(\vartheta)$ of a theory $\vartheta = \vartheta(O_1, . . ., O_m; T_1, . . . , T_n)$ with observational predicates $O_1, . . . , O_m$ and theoretical predicates $T_1, . . ., T_n$ is just the result

$$\exists X_1 . . . \exists X_n \vartheta(O_1, . . ., O_m; X_1, . . ., X_n)$$

of existentially quantifying away the theoretical terms $T_1, . . ., T_n$ and replacing them by new variables $X_1, . . ., X_n$ of the appropriate arity and type (or sort, if the underlying logic of $R(\vartheta)$ is taken to be first-order). The replacement of ϑ by $R(\vartheta)$ preserves the class of derivable consequences involving the observational vocabulary – what Ramsey called "the system of primary propositions" of ϑ; it is in this sense that, for Ramsey, $R(\vartheta)$ can

---

[8] See Cartwright (1987, p. 117) for an elaboration of this point.
[9] Ramsey (1929). References to "Theories" are to its publication in Braithwaite (1931); to find the corresponding page number for the paper's reprinting in Mellor (1990), simply subtract 100.

be said to capture the content of ϑ.[10] But of course $R(\vartheta)$ must necessarily diverge from ϑ in those consequences involving the theoretical vocabulary (Ramsey's so-called "secondary propositions" or "secondary system").

Now the point of Ramsey's proposal is to address the role of theoretical terms in the deductive structure of ϑ, and then to address their role in those deductions that are relevant to the derivation of *primary* propositions. Ramsey in effect observed that if ϑ contains sufficiently explicit deductions of its primary propositions then these deductions have a representation using $R(\vartheta)$. But since in $R(\vartheta)$ the "propositions" which are the transforms of secondary propositions contain variables wherever the original propositions contained theoretical terms, their meaning is exhausted by the contribution of the observational vocabulary they contain. As Ramsey put it,

> So far ... as *reasoning* is concerned, that the [transforms in $R(\vartheta)$ of the secondary propositions] are not complete propositions makes no difference, provided we interpret all logical combinations as taking place within the scope of a [single existential] prefix ... For we can reason about the characters in a story just as well as if they were really identified, provided we don't take part of what we say as about one story, part about another.

> We can say, therefore, that the incompleteness of the "propositions" of the secondary system affects our *disputes* but not our *reasoning*. ("Theories," p. 232)

And since, for Ramsey, it is only our reasoning we need to reconstruct, this incompleteness is irrelevant. The idea of a Ramsey sentence depends on nothing more contentious than this elementary observation about the formal character of logical derivation. But of course to grant this is not to concede the correctness of the view of "secondary propositions" – as mere auxiliaries in the derivation of primary propositions – which it advances.

Regarding Ramsey's remark about disputes versus reasoning, it is obvious that if two theories T_1 and T_2 conflict in their secondary propositions, this conflict need not be preserved under "ramsification" but might well be existentially generalized away. It is, however, possible to take things a step further by recalling an observation of Jane English. It is a consequence of the Craig interpolation theorem that if two first-order theories with disjoint T-vocabularies but coincident O-vocabularies are inconsistent with one

---

[10] The idea that $R(\vartheta)$ isolates the factual content of ϑ is indefensible unless *some* predicates of ϑ are exempted from ramsification; otherwise the "Ramsey sentence" of any consistent first-order theory will be true in all models (L-true), true in every model whose domain of individuals has cardinality n, for some finite cardinal n, or true in every model whose domain of individuals is denumerably infinite. This follows from the fact that if no predicates of ϑ are exempted from ramsification, then the models for the language of $R(\vartheta)$ are all of form $M = \langle M, P(M^1), \ldots, P(M^n) \rangle$, where $P(M^i)$ is the power set of the i-th Cartesian product of M, and a simple application of the observation of Newman discussed below. I am indebted to Aldo Antonelli for suggesting this remark.

another, then there is a sentence σ in their common O-vocabulary which *separates* them, i.e. which is such that T_1 implies σ while T_2 implies $\neg\sigma$. Hence T_1 and T_2 cannot be compatible with the same "data." The proof is a straightforward application of Craig interpolation and the compactness theorem: if $T_1 \cup T_2$ is inconsistent, then by Compactness there are finite subsets $\Sigma_i \subseteq T_i$ ($i = 1, 2$) such that $\Sigma_1 \cup \Sigma_2$ is inconsistent. Let σ_i be the conjunction of the sentences in Σ_i. Then σ_1 implies $\neg\sigma_2$. Hence by Craig Interpolation, there is a sentence σ such that σ_1 implies σ and σ implies $\neg\sigma_2$, where σ is in the common vocabulary of σ_1 and σ_2, i.e. σ is an O-sentence formulated in the common observational vocabulary of T_1 and T_2. But then T_1 and T_2 cannot both be compatible with all observations: if σ holds, then T_2 is false, and if it does not hold, T_1 is false.[11]

We can turn this observation about first-order theories into one about Ramsey sentences: there cannot be two incompatible Ramsey sentences both of which are compatible with all the true O-sentences. Provided our original theories are first-order, it does not matter for the application of first-order model theory that their Ramsey sentences are second-order. The essential idea behind the use of the Ramsey sentence, as we have noted, is that only the logical category of the theoretical terms is relevant to their role in deducing the primary propositions. So long as the replacement preserves the logical category of the original vocabulary items, someone who holds that the Ramsey sentence of a theory captures its content can have no objection to a uniform replacement of the theoretical vocabularies of T_1 and T_2 with new nonlogical constants, making their theoretical vocabularies disjoint, and therefore ensuring that T_1 and T_2 satisfy the hypothesis of our corollary to Craig Interpolation. The identification of a theory with its Ramsey sentence therefore comes at a price: the notion that two theories might be compatible with the same data and yet conflict with one another must be given up, so that when theories are identified with their Ramsey sentences, a conflict in some secondary proposition must be reflected in a primary proposition.

The general methodological issues addressed by the notion of a Ramsey sentence do not exhaust Ramsey's interest in the elimination of "superfluous elements." A fragment found with "Theories" shows Ramsey to have had an interest in particular cases that is worth noting. It is unfortunate that

---

[11] The discussion in English (1973) is slightly misleading: contrary to what English claims, the argument on which her observation depends does not use the Robinson consistency theorem but uses – as the argument given in the text shows – a part of a well-known derivation of Robinson's theorem from the theorem of Craig.

this fragment has not been reprinted in any of the published editions of Ramsey's papers, since it shows Ramsey to have perceived with remarkable clarity the foundational problem of characterizing Newtonian space-time.[12] The text is worth quoting in full:

> In a completely satisfactory theory I think we should
> (a) have a complete dictionary
> (b) have no superfluous elements.
> (b) cannot be exactly defined: it means that we cannot get a simpler equivalent theory. But we may be able to do so by a little transformation when we cannot by simply leaving out a part as it stands. Weyl's requirements (1927, p. 87) are *Einstimmigkeit* [unanimity] and no *überflüssen gestandteile* [superfluity of expression].
> Which seem to mean that every theoretical quantity can in principle be evaluated and that all ways of evaluating it lead to the same result. In principle must here mean merely that certain possible courses of experiences would determine its value.
> If not, of course, there is something superfluous[. E.]g. our velocity in absolute space could not be determined, and so some truth-possibilities of theoretical functions give [an] equivalent theory. [Therefore] some economy ought to be possible, but it is not clear how without a good deal of thought. That makes indeed a good exercise.
> What is the proper form of Newtonian Mechanics? Which gives absolute acceleration a meaning[,] absolute velocity[,] none.
> It must be a sort of geometry containing straight lines and a fixed direction. One must give an axiomatic description of such a geometry.[13]

The problem Ramsey has posed – the formulation of a four-dimensional affine geometry for Newtonian mechanics – differs from the general methodological issue addressed by the Ramsey sentence in a crucial respect. The transformation of Newtonian mechanics to R(Newtonian mechanics) takes for granted the theoretical formulation of its secondary system of propositions. But the idea underlying a reformulation of Newtonian mechanics without absolute space is that it leads to a refinement of the theory by eliminating all those *primary* propositions which purport to express a measurement of absolute velocity. By contrast, when we "ramsify" we assume that the *primary* propositions have been correctly circumscribed, since there is nothing in the transformation to the Ramsey sentence capable of correcting matters if this is not the case. Thus, Ramsey's distinction between primary

[12] The issue received its definitive formulation in the philosophy of science literature with the publication of Stein (1967).
[13] Dated August 1929. Frank Plumpton Ramsey Papers, 1920–1930, ASP.1983.01, Archives of Scientific Philosophy, Special Collections Department, University of Pittsburgh. Quoted with permission.

and secondary propositions does not address the kind of "superfluity" that attaches to absolute space and absolute velocity in Newtonian mechanics. The source of their superfluousness is the dynamical structure of the theory: in a theory with a different dynamical structure, absolute space and absolute velocity might not be superfluous, and the propositions reporting their state would properly occur among the theory's primary propositions.

In addition to whatever other criteria they fulfill, in "Theories" Ramsey assumed that the primary propositions of a properly formulated theory satisfy Russell's fundamental principle – indeed, he even goes so far as to say that names of experiences are descriptions unless the experiences named are *present* experiences (see "Theories," p. 213). The epistemic significance of names versus variables is therefore much the same for Ramsey as it is for Russell: names require an account of how they are understood, variables do not. Certainly, Ramsey supposes that our understanding of the vocabulary of the primary propositions (the observation or O-vocabulary) is unproblematic. And since *only* the O-vocabulary is regarded as unproblematic, Ramsey's account is naturally viewed as an extension to the case of theoretical predicates of Russell's analysis of the meaning of names for things falling outside our perceptual acquaintance. By eliminating theoretical predicates and replacing them with variables, Ramsey avoids any difficulty their presence might pose for an empiricist theory of propositional understanding.

Ramsey assumes that the referents of the theoretical predicates are not known by acquaintance and that therefore the theoretical predicates are not understood in accordance with the fundamental principle, but he refrains from offering a positive account of the character of our knowledge of theoretical relations along the lines of Russell's structuralism. Ramsey ignores this question and focuses on another: significantly diverging from Russell, Ramsey turns his attention to the collection of propositions which constitute the theory and then considers the effect of identifying the content of the secondary propositions with their consequences in the primary system.[14] The implicit assumption, that only the vocabulary of the primary system is fully contentful, and that our understanding of the secondary propositions consists in our understanding of logic plus the vocabulary belonging to the primary system, is in essence Russell's view. What is novel is the manner in which Ramsey bypasses the issue of the meaning of the individual terms of the theoretical vocabulary by eliminating them in

---

[14] Compare the following passage: "[T]he meaning of a proposition about the external world is what we should ordinarily regard as the *criterion* or *test* of its truth. This suggests that we should define propositions in the secondary system by their criteria in the primary" ("Theories," pp. 222–3).

favor of variables, since the technical device by which this is achieved leads to the idea (found later, as we will see, in Carnap) of expressing the content of the theory *as a whole* by the set of its primary propositions.*

Before proceeding to later developments, let me briefly summarize the salient points of comparison between Ramsey and Russell. First, Ramsey is prepared to preserve the idea that the theoretical or T-vocabulary is referential, without, however, necessarily preserving the *order* (in the sense of the ramified hierarchy) of the relations, since the systematic substitution of $\exists$-bound variables for T-vocabulary items requires only that the matrix $\vartheta(O_1,\ldots,O_m; T_1,\ldots,T_n)$ be satisfied by entities of the appropriate *simple* type. This is how I understand Ramsey's remark ("Theories," p. 231) that "it is evident that [the values of the bound variables] are to be taken purely extensionally. Their extensions may be filled with intensions or not, but this is irrelevant to what can be deduced in the primary system." Secondly, Ramsey does not claim to be giving explicit or implicit definitions of the vocabulary of the secondary system. (The concept of *implicit definition* is not discussed by Ramsey.) The paper canvasses various possible general strategies for explicitly defining the secondary vocabulary in terms of the primary vocabulary but concludes that this is unlikely since the "secondary system has a higher multiplicity, i.e. more degrees of freedom than the primary ... and such an increase of multiplicity is ... [likely] a *universal* characteristic of useful theories" ("Theories," p. 222). Correlative to this last point, by not offering explicit definitions, Ramsey's approach echoes Russell's contextual definition of problematic names. But there is no question of contextually analyzing the T-terms away in Russell's sense, since $R(\vartheta)$ is only "weakly" equivalent to ϑ – equivalent over the sentences of the primary system. By contrast, for Russell, the transform $[S(n)]$ of a sentence $S(n)$ effected by the description theory of names is supposed to be equivalent to the untransformed sentence. Finally, notice that our comparison of Ramsey and Russell is based entirely on Russell's application of his extension of his theory of descriptions to names of things falling outside our acquaintance – on aspects of the theory of propositional understanding which emerged from his theory of descriptions and description theory of names – and has not at any point appealed to Ramsey's view regarding the *nature* of our knowledge of matter and the theoretical relations of physics, beyond the minimalist claim that such knowledge as we have is not knowledge by acquaintance. It might seem therefore that Ramsey's development

---

* *Note added in 2012.* Regarding this idea of Ramsey's and Carnap's, see especially the discussion in Section 7.2.

of the theory of propositional understanding is viable in a way that Russell's elaboration of it in terms of his structuralism was seen not to be. In the next section, we will see that this is not the case, even under a refinement of the theory introduced by Carnap.

6.4 CARNAP'S RECONSTRUCTION OF THE LANGUAGE OF SCIENCE AND AN OBSERVATION OF NEWMAN

The next major development in the tradition I have been reviewing is Carnap's mature reconstruction of the "language of science," first presented in his Santa Barbara Lecture of 1959,[15] and subsequently developed in Carnap (1963b), (1966), and (1995). Carnap's primary goal was to provide a reconstruction which clearly separated the factual from the nonfactual assumptions of a physical theory.[16] Carnap frequently emphasized that virtually every *unreconstructed* sentence of the language of science and everyday life has both factual and nonfactual aspects, so that a sharp separation of our language into factual and nonfactual sentences is meaningful only relative to a *reconstruction* of that language. Carnap's proposed reconstruction is very simple. Like Ramsey, Carnap focused on the factual content of a theory as a whole. And, like Ramsey, Carnap took this to be carried by its Ramsey sentence.

For Carnap, as for Ramsey, the decisive consideration that justifies locating the factual content of ϑ in $R(\vartheta)$ is that $R(\vartheta)$ has the same O-consequences as ϑ. Carnap, however, seeks to effect a general procedure for isolating the stipulational or nonfactual component of ϑ to a proper part of the reconstruction, with the factual content of the theory being exhausted by its Ramsey sentence. On Carnap's reconstruction, the nonfactual or analytic component is given by what has come to be called the *Carnap sentence*, $C(\vartheta)$ $(= R(\vartheta)$ only if ϑ), of the theory, a sentence which expresses the thought that if anything satisfies the matrix of the Ramsey sentence of the theory, then the referents of the terms of the unramsified theory do.[†] Carnap argues that

---

[15] Edited by Stathis Psillos and published together with an informative introduction as Psillos (2000). For additional historical background, see Psillos (1999, Chapter 3).

[16] Carnap's discussion of reduction sentences and Ramsey sentences in Carnap (1963b, sect. 24) shows this to have been a feature of early formulations of his views.

[†] *Note added in 2012*: This formulation is misleading: it fails to emphasize that the Carnap sentence is concerned with theories that are reconstructed as only *partially interpreted*. (The notion of partial interpretation is explained later in this section.) The Carnap sentence asserts of such a theory that it is true if its Ramsey sentence is true. The sentence therefore has the methodological character of an *implicit definition* of the theoretical terms of the theory, a definition which satisfies a restricted form of the noncreativity condition of the classical theory of definition. For a discussion of the logical properties of the Carnap sentence see Demopoulos (2007).

since $C(\vartheta)$ has the property that $R(C(\vartheta))$ is L-true, and therefore has no O-consequences except those that are L-true, it is nonfactual.

There are a number of simple connections between ϑ, $R(\vartheta)$, and $C(\vartheta)$ which we should record. ϑ is obviously equivalent to the conjunction of $C(\vartheta)$ and $R(\vartheta)$. Under the assumption that the Ramsey sentence $R(\vartheta)$ of ϑ is true,

$$\vartheta \Leftrightarrow R(\vartheta) \text{ only if } \vartheta,$$

i.e. under the "factual" hypothesis that *something* satisfies ϑ, ϑ is equivalent to its Carnap sentence. More significantly,[17] under the hypothesis of the Carnap sentence, it follows that

$$\vartheta \Leftrightarrow R(\vartheta),$$

i.e. it follows that ϑ is *equivalent* to its Ramsey sentence – not just equivalent over the primary system, to use Ramsey's terminology. But since the Carnap sentence is an *analytic*, and hence *necessary*, truth, we may simply say that ϑ is equivalent to its Ramsey sentence *without qualification*. Intuitively, by accepting the Carnap sentence as analytic we exclude as conceptually possible all those models in which the Carnap sentence fails; a model in which $R(\vartheta)$ holds but ϑ fails is simply not a possible model.

As simple and elegant as Carnap's proposed reconstruction is, I think it cannot be accepted as an accurate reflection of our pre-analytic under-standing of our theoretical knowledge about the physical world. That the attempt to do so leads to unacceptable consequences is, I think, the lesson to be drawn from an old observation of Newman (1928), an observation that lies at the basis of the formulation of the earliest and simplest of Hilary Putnam's "model-theoretic arguments."[18] The application of Newman's observation I will be developing – by contrast with Putnam's deployment of his argument against "metaphysical realism" – is completely straightfor-ward. The presentation differs from Newman's only in the use of model-theoretic terminology; conceptually, the point is entirely the same.

Suppose we are given a theory ϑ, all of whose observational consequences are true; it follows from this supposition that ϑ is empirically adequate and consistent. Suppose also that the observational consequences of ϑ can be characterized as a subset of the sentences generated from a given O-vocabulary. Suppose further that the interpretation of the language $L(\vartheta)$ of ϑ is specified *only* for its O-vocabulary, and that the interpretation of the T-vocabulary is fixed only up to the logical category of the T-terms;

---

[17] Cf. Winnie (1970, p. 294).   [18] First presented in Putnam (1977).

$L(\vartheta)$ is said to be *partially interpreted*. Notice that partial interpretation is the reflection in the theory of theories of what, in our discussion of Russell's theory of our knowledge of matter, we identified as his tacit assumption (iv): the logically primitive predicate expressions we are entitled to suppose we understand are restricted to those whose extensions consist of objects of perceptual acquaintance. Without this assumption, the epistemological motivation for distinguishing between the O- and T-vocabularies – and therefore, Ramsey's primary and secondary systems – would be lost.[19] Notice also that it is no objection to the distinction to observe that it cannot be satisfactorily drawn within the *un*reconstructed vocabulary of the language of science. The point of the reconstruction of theories we are exploring is to show that it is possible to achieve the theoretical knowledge we take ourselves to have by showing how – within a reconstruction in which the distinction between the observational and the theoretical does its intended work – we can give a faithful representation of what we take ourselves to know. Provided we can isolate the observable part of any intended model of ϑ, it is always possible to introduce into the reconstructed language of ϑ properly observational predicates – predicates defined in terms of the restrictions of the interpretation of the predicates of the unreconstructed vocabulary to the subset of observable elements of the domain. (Indeed, as we will see later, this possibility establishes the connecting link between Carnap–Ramsey and "constructive empiricism.")

It is clear from the foregoing that for Carnap ϑ is effectively identified with the matrix of its Ramsey sentence and that this is entirely in keeping with his view, since, of the Ramsey sentence of ϑ, Carnap says that while it

---

[19] The well-known paper of Lewis (1970) is often represented as a part of the Carnap–Ramsey program we have been reviewing. This assimilation is a mistake. Aside from Lewis's adoption of the formal apparatus of the Ramsey and Carnap sentences, both his positive contribution and the philosophical concerns he addresses are orthogonal to Carnap–Ramsey. Lewis is himself explicit on the central point, namely that his O-vocabulary is not restricted in the way it must be for the Carnap–Ramsey reconstruction to have the epistemological interest claimed for the O-vocabulary by that reconstruction. A close study of Lewis's paper shows that its contribution is wholly logical: Lewis assumes that the theories to which his analysis applies are such that their O-terms implicitly define the T-terms in the sense that every automorphism of a model for the language of the theory that preserves the relations denoted by the O-terms will also preserve the relations denoted by the T-terms. Lewis's principal contribution consists in imbedding this assumption and the formal features of the account of Ramsey and Carnap into a framework whose underlying logic allows for the possibility of denotationless terms. While this is not without an interest of its own, it lacks the philosophical motivation that prompts the Carnap–Ramsey reconstruction. *Added in 2012*: It is possible to show that the way in which Lewis's O-terms – especially his "O-mixed terms" – restrict the interpretation of his T-terms is essential to his account and to its principal difference from the Carnap–Ramsey reconstruction. However, I would not now characterize this as the assumption that the O-terms of a theory *implicitly define* its T-terms. The point also bears on the comparison of Lewis with Winnie (1967). However the discussion of these matters requires more space than can reasonably be accommodated by a footnote.

does indeed refer to theoretical entities by the use of abstract variables . . . it should be noted that these entities are . . . purely logico-mathematical entities, e.g., natural numbers, classes of such, classes of classes, etc. Nevertheless, $R(\vartheta)$ is obviously a factual sentence. It says that the observable events in the world are such that there are numbers, classes of such, etc., which are correlated with the events in a prescribed way and which have among themselves certain relations; and this assertion is clearly a factual statement about the world. (1963b, p. 963)

That is to say, Carnap shares with Ramsey the idea that the "factual content" of the theory consists in its consequences in the language of the primary system; the theoretical claims taken by themselves are "purely logico-mathematical" in character. It has been objected[20] that passing to the Ramsey sentence of a theory that is advanced as merely an idealization, such as the theory of ideal gases, we lose the intuitive content of the original theory, since the Ramsey sentence represents it as advancing the false claim that, for example, *there is* an ideal gas, contrary to our pre-analytic understanding of the theory of ideal gases. However, as our quote from Carnap shows, the understanding of the distinction between theories involving *idealization* and theories involving *abstraction* on which this pre-analytic intuition depends is precisely what his use of the Ramsey sentence rejects. It cannot be emphasized too strongly that for Carnap all that needs to be preserved is reference to entities of the appropriate logical category, and it is a matter of indifference whether these entities are "concrete" or "ideal" or "abstract." Although the case of idealization points to a difficulty in the equation of the factual content of a theory with the content of its Ramsey sentence, the difficulty goes much deeper than the failure of this equation to capture our pre-analytic understanding of theories of ideal systems. This is what we will now show.

Since ϑ is consistent, there is an abstract model M of the T-sentences of ϑ, although nothing of philosophical interest would be lost if we were forced to proceed *directly* from the assumption that ϑ has such an abstract model and were unable to infer this (e.g. because of the higher-order character of the language in which ϑ is formulated) from the mere consistency of ϑ. Without any significant loss of generality or philosophical interest, we may choose an M that is a model of the same cardinality as ϑ's intended domain. Let W be the domain we take ϑ to make assertions about – the intended domain of ϑ. By hypothesis, W has the same cardinality as M. It is therefore possible to extend the partial interpretation to the theoretical vocabulary of ϑ by letting each predicate of its theoretical vocabulary denote the image in W of its interpretation in M under any one–one correspondence between M

---

[20] See English (1973).

and *W*. For example, suppose *T* is a binary theoretical relation in $L(\vartheta)$. Then the interpretation T^W of *T* in *W* is *defined* as the image under φ, φ one–one from *M* on to *W*, of its interpretation T^M in *M*. Since by construction <*a*, *b*> is in T^M if, and only if, <φ*a*, φ*b*> is in T^W, φ is an isomorphism; and therefore, if *M* is a model of ϑ, so is *W*.

Let $\mathfrak{I}$ be the interpretation of ϑ's T-vocabulary in *W* that we have just described. Any theory of knowledge and reference that is incapable of distinguishing truth from truth under $\mathfrak{I}$ is committed to the implication,

$$\vartheta \text{ is true if } \vartheta \text{ is true under } \mathfrak{I}.$$

But modulo our assumption about cardinality, that ϑ is true under $\mathfrak{I}$ is a matter of model theory. $\mathfrak{I}$ is arbitrary, and the construction which employs it is clearly unacceptable, since it trivializes the question whether ϑ is true. Any account of our theoretical knowledge that cannot exclude $\mathfrak{I}$ as an interpretation of the language which adequately captures the interpretation under which we suppose that ϑ is true, cannot account for our naive confidence in the belief that our theories, if true, contain significant theoretical truths about the world. By equating truth with truth under $\mathfrak{I}$ we rob our knowledge of the truth of our theoretical claims of its a posteriori character; modulo a single assumption about cardinality, the theoretical statements of an empirically adequate theory come out true as a matter of metalogic. But we take the truth of ϑ over its intended domain to be a significant a posteriori truth, not one that is ensured by what is virtually a purely logical argument. Since there is nothing in Carnap's reconstruction to exclude the identification of truth with truth under $\mathfrak{I}$, an essential feature of the truth of our theories has been lost, and Ramsey's and Carnap's reconstructions cannot therefore be judged successful.[21]

The conclusion we have just reached was partly anticipated by John Winnie (1967).[22] Winnie's emphasis is, however, different from ours, since for him the

---

[21] Notice that the observation on which Putnam's model-theoretic argument depends is quite elementary. In particular, the argument does *not* appeal to any major metalogical result, let alone anything so sophisticated as the Löwenheim–Skolem theorem. One can formulate things so that the argument appears to require the model-existence lemma, but this is misleading since the semantic consistency of ϑ would certainly be taken for granted by Carnap–Ramsey and therefore hardly needs to be derived. Notice also that, as presented here, the argument is not a permutation argument; such arguments are used by Putnam in other contexts (see the next footnote), but they are not essential to the concerns raised here.

[22] The existence of an arithmetical model is the content of Winnie's second theorem and the main focus of his paper. His first theorem shows that, modulo a trivial restriction (noted below), if *W* is a model of ϑ, so is *W**, where *W** is like *W* except for the interpretation of some T-predicate. Winnie's discussion assumes that the domain of any model of ϑ is the disjoint union of subdomains W_U and W_O of (respectively) unobservable and observable entities, with the interpretation of the T-predicates

salient point is that if we are given a "physical" interpretation under which ϑ comes out true, it is virtually always possible to find another, as it happens abstract, arithmetical, and unintended, interpretation that also makes ϑ true. The problem Winnie emphasizes is therefore one of finding conditions that will exclude such unintended interpretations without compromising the assumptions of the framework within which the reconstruction is expressed. However, this way of presenting the difficulty is misleading. The Carnap–Ramsey reconstruction is unacceptable because it implies that the existence of an *abstract* model for ϑ suffices for its truth over ϑ's *intended* domain. What the reconstruction misses is a pre-analytic intuition that governs our conception of truth, and this is missed even when we set aside any difficulty in fixing ϑ's intended domain.

6.5 EXTENSION OF THE FOREGOING TO CONSTRUCTIVE EMPIRICISM

There is a common misperception of the point of Newman's observation, one for which Putnam is largely responsible. The misperception is that the interest of the model-theoretic argument must stand or fall according to how successfully it refutes "metaphysical realism." Perhaps there is an interesting position of this character which the argument refutes. But to show this we would have to engage in a rather involved investigation into the nature of realism, and should we fail to find a plausible version of metaphysical realism to serve as a suitable target for the argument, we might be led to suppose that the observation on which it is based fails to show anything of philosophical interest.[23] Nothing could be further from the truth. In addition to its relevance to the Carnap–Ramsey reconstruction just reviewed, there is an "epistemology of science" to which Newman's observation applies virtually directly.

Bas van Fraassen's constructive empiricism is essentially characterized by a central tenet and a pair of definitions. The *central tenet* is that it is always more rational to accept a theory as empirically adequate than to believe it true. The *definitions* are: (i) to *accept a theory as empirically adequate* is to hold that "the phenomena" form a substructure of a model belonging to the class of models that *is* the theory; and (ii) to *believe a theory true* is to hold

---

restricted to W_U. The trivial restriction is that if a T-predicate T is monadic, there is a u in W_U such that u is not in the interpretation T^W of T, and if T is n-adic, there is a u in W_U such that u is not a component of some n-tuple in T^W. Putnam (1981, p. 217) appeals to a permutation argument of this sort, without, however, correctly identifying the conditions under which the argument is valid. For a counterexample to Putnam's claim see Keenan (2001, sect. 1).

[23] See, for example, Chambers (2000).

that it contains the world among its models. What distinguishes constructive empiricism from its less circumspect and nonempiricist opposition is its agnosticism regarding truth and the fact that the distinction between what is and is not observable is not drawn on the basis of vocabulary, but concerns the demarcation of substructures of the models which comprise the theory. Constructive empiricism does not deny that theories – i.e. the models theories comprise – contain unobservables in their intended domains; it is merely agnostic regarding a theory's claims about them. What it holds in common with Carnap–Ramsey – when formulations are transposed to the framework preferred by the semantic view of theories – is the notion that to assert a theory as true is to say that *there is a model* belonging to the theory (recall that for van Fraassen a theory is identified with a class of models) that "corresponds" to the world:

> My view is that physical theories do indeed describe much more than what is observable, but that what matters is empirical adequacy, and not the truth or falsity of how they go beyond observable phenomena . . . To present a theory is to specify a family of structures, its *models*; and secondly, to specify certain parts of those models (the empirical substructures) as candidates for the direct representation of observable phenomena. The structures which can be described in experimental and measurement reports we can call *appearances*; the theory is empirically adequate if it has some model such that all appearances are isomorphic to empirical substructures of that model. (van Fraassen 1980, p. 64)

But so long as a theory consists of empirically adequate models with domains of the right cardinality, constructive empiricism is committed to the view that a theory is true provided that it is empirically adequate. The issue is not constructive empiricism versus realism but whether the framework within which constructive empiricism is expressed – the semantic view of theories – successfully recovers the intuitive sense we attach to ascriptions of truth to our theoretical claims. Constructive empiricism supposes that it has captured this sense, and proceeds to oppose the belief that our theories are true. The difficulty is that on its own account of the truth of theoretical claims that go beyond the phenomena there is virtually nothing to choose between truth and empirical adequacy. This is contrary to the idea – evident to pre-analytic intuition and, apparently, to constructive empiricists – that it is a substantial philosophical commitment to believe a theory true rather than to accept it as merely empirically adequate. What is in dispute is not whether we should believe our theories true – although this is of course an important issue – but whether constructive empiricists have captured what such a belief consists in, since, using our earlier construction, we can always ensure that any theory that saves the phenomena contains the world among the models it comprises: by hypothesis, one of the theory's empirically adequate and partially abstract models is in one–one correspondence with the world or "intended domain" of observable and unobservable entities. In exact analogy with our argument against Carnap–Ramsey, *define* the theoretical relations on this intended domain so that, for example, T^W holds of <a, b> if and only if <$\varphi^{-1}a$, $\varphi^{-1}b$> belongs to T^M, where φ is one–one from M on to W and is the identity map on $M\cap W$ – the observable part of M and of W. This transforms a theory that merely saves the phenomena into a true one, since it ensures that the world is a member of the class of models that is the theory. But then what has happened to the central tenet of constructive empiricism, according to which it is always more rational to accept a theory as empirically adequate than it is to believe it to be true? This appears to be based on nothing more than an agnosticism regarding the cardinality of the intended domain. Our deployment of Newman's observation thus shows that the combination, constructive empiricism plus semantic view of theories, yields an inherently unstable position.

6.6 PUTNAM'S MODEL-THEORETIC ARGUMENT AND THE SEMANTIC VIEW OF THEORIES

It might seem as if the considerations we have reviewed derive their force against Carnap because of his commitment to what van Fraassen (1989, Chapter 8, sect. 6 and Chapter 9, sect. 3) has called the "syntactic" view of theories, in contrast to van Fraassen's own, "semantic" view of theories. Recall that on the semantic view, we explicitly assume that

$$\vartheta \Longleftrightarrow \text{ is true} \Longleftrightarrow \vartheta \text{ has a class of } \textit{intended} \text{ models},$$

so that on this view the truth of ϑ is always understood relative to an intended model. Indeed, on the semantic view a theory is *identified* with a class of intended models, and the truth of a theory consists in the world being one among the class of intended models which *is* the theory. It might be thought that the restriction to intended models, on which the semantic view insists, renders it not vulnerable to the difficulties to which Carnap's syntactic account is subject.

There are, however, *two* senses of 'intended model' (or 'intended interpretation') at play here. The first sense is related to the distinction between what is sometimes called "real" second-order logic and the logical systems considered by Leon Henkin in his proofs of completeness. Such systems allow for general or Henkin models and this leads to a distinction between theories which hinges on the scope of the domain of the property and relation variables of a theory. This is the sense of 'intended model' that proponents of the semantic view emphasize when expounding their position, since unless a theory is identified with a class of intended models in this sense – thus invoking in an essential way a notion of theory that is not purely formal (but is what is sometimes called "semi-formal") – the semantic approach's distinctive emphasis on *mathematical* as opposed to *meta-mathematical* methodology cannot be sustained. The semantic approach correctly observes that the use of axiomatics in foundational studies in the exact sciences is rather different from its use in metalogical investigations. Roughly speaking, in the former case we are concerned to characterize some property or other – say, the property of being a Euclidean space – and we proceed to do so by capturing (up to isomorphism) a class of structures which exhibit the desired property. In the latter case, our interest in axiomatization is motivated by an altogether different set of considerations, such as, for example, our interest in the recursive enumerability of a particular set of truths.

Interesting as this methodological observation is, the sense of 'intended model' that needs to be addressed in order to avoid Putnam's argument is what is involved when, in describing the meaning of some fragment of our language, we distinguish one interpretation of the nonlogical constants of this fragment as the intended one. This problem is at best only partially addressed by the issues peculiar to the interpretation of the variables of the underlying logic; while this fixes the domain from which the properties and relations are drawn, it leaves unsettled *which* relations on the domain are the subject matter of the theory. Putnam's observation, that we cannot take the intended model in this second sense to be singled out by $R(\vartheta)$, also applies to the semantic view's identification of a theory with a class of intended models, since this identification concerns only the first sense of 'intended model'. To address the second sense of 'intended model', involving as it does the constant terms, would be to engage in an analysis of the meaning of the language of ϑ, contrary to the promise of the semantic view that it allows us to dispense with such questions altogether. Invoking the semantic view appears therefore to have brought us no closer to a satisfactory account of our theoretical knowledge.

Van Fraassen (1997) addresses Putnam's model-theoretic argument at some length, without, however, explicitly observing its bearing on the formulation of the constructive empiricist theory of theories. The core of his analysis is that Putnam has exposed a new "pragmatic paradox," a paradox that emerges when we question the interpretation of our own language.[24] Van Fraassen's idea is that while we can coherently ask of a language we do not understand whether a predicate of that language refers to (say) green things, we cannot coherently ask of a language we *do* understand whether 'green' refers to green things, since it is implicit in the idea that 'green' is a predicate of *our* language that we know it to refer to green things. The syntactic view of theories falls victim to Putnam's argument because it seeks to resolve the problem of fixing an interpretation of $L(\vartheta)$ by purely formal means, something which the model-theoretic argument shows particularly clearly to be a mistake. By contrast, although the semantic view does not address the question of how the interpretation of $L(\vartheta)$ is fixed, this is not a defect because the correctness of the intended interpretation is a presupposition to which the semantic view is entitled – it is a presupposition of the pragmatic background within which the models that *are* the theory are given and within which the semantic view is expressed.

Here I think it is important to be clear on the kind of epistemological question Putnam's argument raises regarding the intended interpretation of $L(\vartheta)$. We may distinguish a skeptical interpretation of the question, 'How do we know that "green" refers to green things?' from an entirely different and nonskeptical interpretation of this question, one according to which we presume that we know that 'green' refers to green things and ask for an account of our knowledge of this fact. Clearly, under the latter interpretation, this is a question we can ask about *any* language – including languages we ourselves understand – without raising any hint of paradox, pragmatic or otherwise. Even if it is not legitimate to question whether the reference of one's language is correctly fixed in accordance with its customary interpretation, it is surely legitimate to ask *how* it is fixed. Carnap–Ramsey attempts to address this question by arguing for the adequacy of a reconstruction in which the theoretical vocabulary is eliminated; if such a reconstruction were successful, the question of how the reference of the theoretical vocabulary is fixed could be put to one side. What is peculiar about the Carnap–Ramsey reconstruction is that it articulates its empiricist commitments by isolating the terms of the theoretical vocabulary as those for which an explanation of reference is especially pressing. Even if we believe that Carnap–Ramsey is insufficiently critical of its account of how the reference of the observational vocabulary is fixed, it is nevertheless intelligible that it should be puzzled by how the reference of the theoretical vocabulary might be fixed: the

---

[24] Putnam makes a similar observation in his original formulation of the argument in order to motivate his "internal realism." For van Fraassen the observation signals a move from metaphysics to pragmatics.

empiricist commitments it shares with Russell's theory of knowledge and propositional understanding yield a model for addressing the reference of the observational vocabulary that has no obvious application to the case of theoretical vocabulary. The problem thus becomes one of explaining – or explaining away – the reference of this vocabulary; hence the Carnap–Ramsey strategy just reviewed. But the semantic view simply fails altogether to address the problem.

6.7 THE PROBLEM CLARIFIED AND RESOLVED

To be significant, the claim that ϑ's theoretical sentences are true must be understood relative to an interpretation of the language of ϑ that is capable of being given independently of the truth of the theoretical sentences themselves. Pre-analytically, we assume that the intended interpretation of the theoretical vocabulary is such an interpretation and that truth with respect to it is not something that can be settled by a purely logical argument. While the rational reconstructions we have been considering are perfectly adequate accounts of how we reason with theories, they are unable to accommodate this feature of our ascriptions of truth. This is the central lesson of Newman's observation.

In order to understand what has gone wrong, and to better see the nature of the problem we have uncovered, it is instructive to look at Russell's formulation of what he calls "the problem of interpretation." The problem and its significance are explained in the introduction to *The Analysis of Matter*.

It frequently happens that we have a deductive mathematical system, starting from hypotheses concerning undefined objects, and that we have reason to believe that there are objects fulfilling these hypotheses, although, initially, we are unable to point out any such objects with certainty. Usually, in such cases, although many different sets of objects are abstractly available as fulfilling the hypotheses, there is one such set which is much more important than the others . . . The substitution of such a set for the undefined objects is "interpretation." This process is essential in discovering the philosophical import of physics. (*AM*, pp. 4–5)

For Russell, the point-instants of the theory of space-time pose a problem exactly similar to that posed by the numbers in Peano's axiomatization of arithmetic. Recall that so far as the Peano axioms are concerned, *any* ω-sequence can form the basis of a suitable model of the axioms. But among ω-sequences, there is one that is distinguished, namely the one which consists of "the" cardinal numbers, since, as Russell says, this fulfills the requirement "that our numbers should have a *definite* meaning, not merely that they should have certain formal properties. This definite meaning is defined by the logical theory of arithmetic" (Russell 1919, p. 10). The "logical theory of arithmetic" associates 'o' with the class of all null classes, '1' with the class of all singletons, '2' with the class of all couples, etc. Although any association with the members of a progression will succeed in giving a "definite meaning" to the numeral names, Russell, like Frege before him, argued that among the various possible definite meanings, the one indicated is distinguished by the fact that it captures our use of the numeral names in our judgments of cardinality. The Frege–Russell cardinals are perhaps the simplest example of a successful application of the method of logical construction to a problem of interpretation in Russell's sense. We would today express this by saying that the set-theoretic construction consisting of the Frege–Russell cardinals provides a *representation* of any model of the Peano axioms. The abstractness of the number-theoretic axioms consists in the fact that they fail to distinguish, among all possible ω-sequences, the one which is associated with their foremost application. This is what the logical theory is held to achieve.

The way to think of Russell's construction of point-instants is to see it as an attempt to accomplish for the theory of space-time what the definition of the Frege–Russell cardinals achieved for the theory of the natural numbers. In each case, the axiomatically primitive notions of number and point-instant are replaced by something else – classes of equinumerous classes and maximal copunctual classes of events, respectively – in order to display the canonical applications of the theories in which these notions occur.[25] In the

---

[25] To fully understand the program of logical construction advanced in *The Analysis of Matter*, it is necessary to look in detail at the notion of copunctuality and the "logical constructions" the book advances. A complete study would be a large undertaking, but the basic idea may perhaps be at least indicated: a class of events is said to be *copunctual* when every five-tuple or quintet of events in the class have a "common overlap." When a quintet of events have a common overlap, Russell says that the five events stand in the relation of *copunctuality* (*AM*, p. 299). The terminology is potentially confusing since the predicate, '*x* is copunctual,' refers to a *property of classes of events*, while *copunctuality* is a five-place *relation between events*. The use of quintets rather than some other number of events arises for technical reasons having to do with the dimensionality of the space whose points (point-instants) are being characterized as logical constructions – set-theoretic structures, as we would say today – out of events. Russell's point-instants are defined as maximal copunctual classes of events – maximal, that is, with respect to class inclusion. Assuming that there are copunctual classes of events, the success of the proof of the existence of space-time points, which occupies Chapter xxviii, turns on showing that every such copunctual class can be extended to a maximally copunctual class. Russell's proof uses Well Ordering, applied to the domain of all events, and his theorem has an evident similarity to the ultrafilter (or maximal dual ideal) theorem for Boolean algebras, with the property of being a copunctual class playing a role analogous to that played by the finite intersection property in the context of the representation theory of Boolean algebras. Indeed, the analogy can be developed further to illuminate the difference between Russell's construction and the earlier, but more restricted, construction by Whitehead in terms of "enclosure

arithmetical case, the canonical application of the theory was the use of numbers as cardinal numbers. Under the influence of Eddington,[26] Russell took the canonical application of the theory of space-time to be its use in *measurement*, a view to which both of them were led by the fundamental role of length and time in the measurement of physical magnitudes. Since Russell's construction of point-instants in terms of events is compatible with the assumption that events comprise only finite volumes, the representation of point-instants by classes of events was motivated by the observation that although there is no bound on the precision we may achieve in any actual measurement, we are always restricted to finite quantities.

There is, however, an important difference between the arithmetical and spatio-temporal cases, one which arises from the fact that they involve applications of theories of very different character. In the context of Russell's logicism, the central question the arithmetical case raises is: 'Given a domain of individuals of the right cardinality, can we recover the structure of the numbers as a theorem of *Principia*?' It is a remarkable and insufficiently appreciated fact that we can. The answer is not obvious since Russell is not simply assuming that the numbers occur among the elements of the domain of individuals, but constructs them at a higher type. Russell calls a class of individuals *inductive* if it belongs to every class which contains the null class of individuals and is closed under the addition of singletons. (Or equivalently, if it is in one–one correspondence with the integers less than n for some integer n. This terminology is evidently motivated by the analogy with the principle of mathematical induction and the definition of the finite numbers.) A class is *noninductive* if it is not inductive. Whitehead and Russell's Axiom of Infinity states that the class of individuals or entities of Type 0 is noninductive. Whitehead and Russell (1912, *124.57) then prove – without the Axiom of Choice, and therefore, one might well argue, by employing only logical modes of reasoning – that the Frege–Russell cardinals, which are entities of Type 2 in the simple type hierarchy, are "reflexive" or *Dedekind*-infinite, and thus form the domain of a model of the Peano axioms.[27]

Now for Russell the analysis of matter just *is* the extension of the method of logical construction to physics in general, and to the theory of space-time

series." Whitehead's construction of spatial points consists in identifying a point with a class of nested volumes, nested by the relation of spatial inclusion. Transposed to the spatial case for comparison, Russell's construction requires only the existence of a common spatial overlap among the volumes belonging to the class of volumes which *are* the point. Russell's notion generalizes Whitehead's enclosure series in the way in which the notion of filter generalizes that of a nested sequence or chain.

[26] See, for example, the discussion in *AM* (pp. 90–4).

[27] See Boolos (1994) for a discussion and exposition of the proof.

in particular. Restricting our attention to the space-time case, here the successful execution of Russell's program and a satisfactory solution to his formulation of the problem of interpretation requires that every abstract model of the theory should have an isomorphic representation by one constructed in terms of maximal copunctual classes of events, where, in analogy with the use of the Axiom of Infinity in the number-theoretic case, events are presumed to comprise a countable collection of concrete individuals (for Russell, events are the "ultimate constituents of the world"). The program of construction requires, quite properly, and again in parallel with the number-theoretic case, that it be provable that the class of events gives rise to an isomorphic representation of any model of the theory of space-time. Thus formulated, the program of logical construction is a familiar part of the nature and methodology of representation theorems, a part which Russell understood perfectly well.

It is important not to mistake the successful execution of the program of logical construction with the vindication of the central epistemological contention of Russell's structuralism: from the fact that the representation of any model of space-time need only be structure-preserving, it might seem to follow that the knowledge expressed by the original theory – in this case, the theory of space-time – is, in the sense appropriate to Russell's theory of knowledge, purely structural. As Russell understands them, the analysis of matter and (more specifically) the problem of the interpretation of the theory of space-time do not in any way require nonstructural knowledge of spatio-temporal relations. This is not an oversight, but an essential feature of the notion of analysis in which Russell is engaged: nonstructural knowledge is not required because the task which the program of interpretation sets itself is the purely mathematical one of constructing an isomorphic representation of the space-time of one or another physical theory, a representation defined over point-instants, appropriately constructed as maximal copunctual classes of events of finite extent.

There are many obstacles to successfully deploying the method of interpretation or logical construction in aid of structuralism. For example, there is a difficulty already at the level of point-instants, since only some of the events used in their construction are the loci of percepts. The construction requires that all of the events that go to make up a point-instant are in the field of the relation of copunctuality. Thus, even the specification of the domain, over which the spatio-temporal structure is to be defined, implicitly presupposes the existence of nonstructural knowledge of the relations among events which are not percepts. When this assumption is dropped, it is not clear whether the definition of the manifold of point-instants can be carried out without the extension of a copunctual class to a maximally copunctual class involving the addition of events which are not percepts.

But the central difficulty arises from the fact that for Russell a successful interpretation of spatio-temporal concepts is not essentially different from the analysis and interpretation of arithmetical concepts with which *Principia* is occupied. Both projects proceed relative to a nonlogical assumption regarding the nature and cardinality of the domain over which the construction is effected, and in both cases the problem is to show that a particular mathematical structure – that of the natural number system or of space-time – can be represented by a logical construction. Such a notion of interpretation addresses one way of understanding the problem of the applicability of a mathematical framework, one that is resolved for Russell once the fundamental objects of the framework are exhibited as set-theoretical constructions – in the case of space-time, when point-instants are constructed from "percepts," or, more generally, from events of finite extent – and appropriate relations are defined over them.

Suppose we allow Russell the assumptions he requires for his construction of point-instants to succeed, and grant that *overlapping* and *copunctuality* are relations which are both perceptible and hold among events which are not percepts. Then even leaving to one side the conflict this poses for his theory of knowledge – for the structure–quality division of what we are capable of knowing regarding the material world – there remains an important difficulty with Russell's notion of interpretation, a difficulty which attaches to the adequacy of his philosophy of physics and theory of knowledge. Russell's approach makes perfect sense when the theory of space-time is conceived as the a priori or quasi-a priori background for spatio-temporal measurement. But from our post-Einsteinian perspective there is more to the theory of space-time than understanding its role in measurement. To accommodate this additional element, our analysis of the interpretation of spatio-temporal concepts must be capable of revealing the theory of space-time as an a posteriori theory of the spatio-temporal structure of the world. *Russell's notion of interpretation as logical construction fails to address this task.* And while it is certainly true that knowledge of space-time structure rests on an interpretive claim, the nature of this interpretive claim is not illuminated by Russell's notion of interpretation. To see this, notice that there are at least three notions of interpretation that are relevant to space-time theories; of these three notions, Russell countenances only the first two.

(i) There is the official notion, the one to which Russell's structuralism entitles him. An interpretation in this sense "constructs" the domain of objects that the theory is normally interpreted as quantifying over. This is what the set-theoretic construction of point-instants seeks to secure in the case of theories of space-time. But not having access to the relations on this domain, providing an interpretation in Russell's sense reduces to the purely mathematical problem of finding some family of relations definable over the constructed objects which constitute the basis for a model of the theory. By its very nature, such an interpretation cannot succeed in illuminating the epistemic status of the theoretical claims of an empirical theory, since there will always be *some* family of relations having the appropriate structure. This, in essence, is just Newman's observation.

(ii) There is a second sense of 'interpretation' that arises if we ignore the constraints of structuralism and allow that we have access to the relations of the theory in addition to having specified the domain of objects among which they hold.[28] The problem of interpretation then consists of two tasks: the purely logical one of showing the sufficiency of a set of postulates regarding the relations of the model – do the postulates capture the basic truths of the original theory? – and the epistemological task of motivating the "naturalness" of the properties the postulates impose. Although interpretation in this sense is not intrinsically incapable of illuminating the epistemic status of theoretical claims – indeed it is hardly different from articulating a well-motivated axiomatization of the theory – its success turns entirely on how it is elaborated in particular cases, on the persuasiveness and sufficiency of the individual axiomatization on offer.

(iii) But there is an altogether different notion of interpretation, one that neither of the previous notions captures. It arises when we expect interpretive claims to clarify the nature and epistemic status of our theoretical commitments in a sense that I will try to clarify by briefly reviewing the analysis that lies at the basis of our knowledge of the space-time structure of special relativity and its divergence from Newtonian space-time.[29]

---

[28] See Anderson (1989) for an authoritative discussion of Russell's investigations along these lines.

[29] The discussion which follows is greatly indebted to DiSalle (2002) to which the reader is referred. *Added in 2012*: The case of simultaneity is better treated above in Chapter 2; see also the remarks regarding criteria of identity and the epistemological difference between applied arithmetic and applied geometry in the introduction to this volume.

Generally speaking, we wish to discover those criteria of application that our use of a theoretical predicate or relation-expression presupposes, so that our assertions regarding the structure in which the associated relation occurs are seen to have the epistemic status that pre-analytically we suppose them to have. It is a presupposition of our successful reference to theoretical relations that we know what the relations are that our theoretical predicates refer to. Even if we take for granted our knowledge of what relation a particular predicate refers to, this falls short of assuming everything of interest that might be asked regarding our knowledge of the reference of the predicates of our language. We have seen that whatever such knowledge consists in, it is not recoverable within the simple empiricist model that Carnap and Ramsey inherited from Russell. The problem is not addressed by a psychological investigation into the origin of our ideas, but requires a conceptual analysis of our theoretical knowledge in general, and of our knowledge of physics in particular, one that clarifies and reveals those assumptions that are implicit in our basic judgments involving the theoretical vocabulary. Assuming that theoretical predicates do refer to their intended referents, we wish to indicate what criteria control the application of such predicates, and we wish to indicate what the conditions of adequacy for such criteria of application might be. The main desideratum for a successful analysis along these lines – and the successful account of our understanding of theoretical predicates such an analysis would yield – is that it should imply that our theoretical claims, when true, are significant truths about the world.

The nature of the problem and the basic idea underlying our proposed solution can be made clear if we take as the predicate whose criterion of application needs to be uncovered the predicate, 'x is simultaneous with y.' This predicate is fundamental to both the Newtonian and Einsteinian cases, since even spatial measurements implicitly assume a comparison of distances at a time, and this requires that we should have settled on a criterion of application for simultaneity. Since simultaneity is an empirical relation among events, one which holds or fails to hold as a matter of fact, we must be able to specify a criterion in accordance with which we can say that pairs of events – including distant events – are in the relation of simultaneity. To grant this is not to say that meaning is verification: it must be part of any view of physics that a presupposition of our use of 'x is simultaneous with y' – both in our theorizing and in the evaluation of our theorizing about space-time – is that it can be shown to conform to an empirically based criterion for applying spatial and temporal predicates. We wish to know those criteria of application that enable us to extend our ordinary judgments involving simultaneity to cases which they may not originally have been designed to cover; and we wish to know whether, in the process, the criteria of application that govern spatio-temporal predicates come to be subject to any new constraints. An interpretation of a space-time theory seeks to address these tasks. Let us consider more specifically what the special relativistic analysis of simultaneity contributes to our understanding of these matters.

Among the criteria of application we actually employ, it is clear that some implicitly assume a process of signaling – as, for example, when we count as simultaneous two distant events which are seen to coincide with some local event, such as the position of the hands of a clock. This is a criterion that is shared not only by the two theoretical frameworks we are discussing – special relativity and Newtonian mechanics – but is also found among our commonsense criteria of application for simultaneity, given the assumption that the signaling criteria employed by our theories are merely refinements of what we deploy in pre-scientific contexts. But because it admits infinitely fast signals, the Newtonian theory allows for a velocity-independent criterion of application for 'x is simultaneous with y.' Within the Newtonian framework, this criterion has the further justification of being, in an appropriate sense, "absolute," since, being independent of any finitely transmitted signal, it is also independent of the relative velocity of the frame of reference, and is therefore formulable without reference to the peculiarities of the circumstances of its application.

There is, however, another criterion of application that is absolute in this same sense even though it is based on the use of signals of finite speed. This is the criterion based on light-signaling that emerges from Maxwell's theory, since according to this theory, the velocity of light is the same for all "observers" – all inertial frames. On Maxwell's theory, a signaling criterion which employs light signals is therefore as frame-independent – as absolute – a criterion of application for 'x is simultaneous with y' as is the Newtonian criterion. But unlike the Newtonian criterion, light-signaling also supports our practical criteria of application for simultaneity in so far as they are based on a finite signaling procedure. It is therefore a criterion that is much closer to what we typically rely upon in our actual judgments regarding spatial and temporal relations; and when light-signaling is understood in the context of Maxwell's theory, it has the added advantage of being as absolute a criterion of application as the Newtonian one.

What is shown by the analysis of simultaneity just sketched is that there is a suitably absolute criterion, that it is based on the common practice of signaling, that it conforms with our most successful theoretical understanding of light transmission, and that under these circumstances it should be acceptable even from a Newtonian perspective. Contrary to a standard understanding of Einstein's methodology, the basis for this choice of criterion of application cannot, therefore, be regarded as founded on nothing more than a free decision. The "choice" is clearly very highly constrained and is a reflection of our pre-analytic practice; what is striking is that it is also a reflection of a methodology that is largely shared by both the Newtonian and the Maxwellian frameworks. But adopting the light-signaling criterion implicit in Maxwell's theory has the unexpected consequence that the relation of simultaneity that it governs is relative – it has the consequence, that is, that "observers" in relative motion to one another will disagree on which events are simultaneous with each other. Most importantly, for our purposes, it has the consequence that, relative to this criterion of application, the space-time which the relation of simultaneity generates is Minkowskian rather than Newtonian. And on this reconstruction, the truth of this structural claim is correctly represented as a significant a posteriori truth about the physical world.

ACKNOWLEDGMENTS

I am particularly indebted to Michael Friedman, Robert DiSalle, and Peter Clark for conversations on the topics of this paper and for their advice on its final form and organization. I also wish to thank David DeVidi, Paul Humphreys, Timothy Kenyon, James Ladyman, Gregory Lavers, Adele Mercier, Christopher Pincock, and Edward Stabler for comments; the Philosophy Departments of Bristol, Queen's, Glasgow, and USC, and the Logic and Philosophy of Science Department of Irvine for the opportunity to present my ideas to them; and the Social Sciences and Humanities Research Council of Canada for financial support. This paper is dedicated to the memory of my friend Graham Solomon, who died on November 1, 2001.

CHAPTER 7

Three views of theoretical knowledge

7.1 INTRODUCTION

The three views of theoretical knowledge that form the subject of my discussion all fall within the framework of the "abstract conception of theories," by which I mean that view of theories which asserts, first, the existence of a partition of the nonlogical vocabulary of the language of a theory into observation and theoretical terms, and holds, secondly, that this partition reflects the fact that our understanding of the theoretical vocabulary is importantly incomplete and problematic in a way that our understanding of the observation vocabulary is not. There are different ways of conceptualizing the incompleteness of our understanding of the theoretical vocabulary. What is characteristic of the views considered here is that for all of them the incompleteness derives from what I will later characterize as "the structuralist thesis," a thesis which implies that the Ramsey sentence of a theory adequately represents the theory's "factual" content. If this consequence of the structuralist thesis could be sustained, then any difficulty the theoretical vocabulary might be thought to pose could safely be ignored; for, given a theory in a language whose nonlogical vocabulary is partitioned into theoretical and observation terms, its Ramsey sentence is the sentence which results when the theoretical terms are existentially generalized away.

The notion of a Ramsey sentence and the abstract conception of a theory which it assumes are sometimes dismissed on the ground that both notions are too far removed from the actual practice of science to have any bearing on even the general and interpretive questions particular theories pose. I reject the view that the abstract conception should be dismissed because of its generality or remoteness from actual practice. Approaches of comparable remoteness have proved invaluable for the philosophical analysis of particular theories, with Hilbert's move toward axiomatics in mathematics the

First published in *British Journal for the Philosophy of Science* 62 (2011): 177–205.

most notable among them. My objection to the abstract conception and its acceptance of the structuralist thesis is specific to that conception and the role of this thesis in it. I do not question the claim that exposing a difference in the conception of the mathematical structure of some aspect of reality often constitutes one of the most fruitful ways of exploring the differences between successive theories.

The objection I will urge against the abstract conception is thoroughly compatible with the idea that "interpretations" of theories are often best viewed as proposals for understanding the nature of the conceptual shift one theory represents by comparison with an earlier theoretical framework; and it is also compatible with the idea that the articulation of the nature of this conceptual shift is in many important cases one that appeals to differences in structure.

From this perspective, Einstein's special theory of relativity is, among other things, an *interpretation* of Maxwell's theory, one that locates the nature of its difference from Newtonian mechanics in the geometry of space-time. This was presaged by Einstein's analysis of simultaneity and the primacy it assigned to the role of light-signaling in the theory of space and time, but it received its fullest expression with Minkowski's formulation of the special theory. This way of understanding the difference between Newton's and Maxwell's theoretical frameworks proved remarkably successful, far overshadowing earlier attempts to provide mechanical models of the aether and dynamical models of the Lorentz contraction. It has also been instrumental in fostering the view that a successful approach to other interpretational problems which arise in the exact sciences will occur only when the appropriate structural change has been isolated and understood.

But my purpose here is not to defend the fruitfulness of general and abstract approaches to our analysis of specific theories and the nature of the theoretical changes they exhibit. Nor is it to stress the importance of structure in any such account. I regard the fruitfulness of abstract approaches and the importance of structure to have long been established. Rather, my concern is with the particular abstract approach to theoretical knowledge that I have called "the abstract conception." My interest in it and its Ramsey sentence elaboration is motivated by their ability to clarify and set in sharp relief different philosophical theories of theoretical knowledge. One philosophical theory the abstract conception and notion of a Ramsey sentence are capable of illuminating is the theory of Russell that is developed at length in *The Analysis of Matter*.[1] Another is the very different theory

---

[1] Russell (1927).

suggested by Ramsey's discussion of the function of theoretical constructs. A third is Carnap's mature rational reconstruction of the language of science, an account in which the notion of a Ramsey sentence figures no less prominently than in Ramsey's own approach, and an account in which the differences between Russell's and Ramsey's views receive a kind of resolution. In so far as Russell, Ramsey, and Carnap are sometimes represented as offering only minor variations on a common view, a novel feature of the following presentation is the nature of the opposition it attributes to Russell and Ramsey and the manner in which it represents Carnap's theory as a resolution of this opposition.

7.2 RAMSEY SENTENCES AND CRAIG TRANSCRIPTIONS

The following discussion of Ramsey sentences is framed in terms of simple arithmetical theories and elementary properties of their models. There is a particularly simple example that illustrates a number of interesting facts about Ramsey sentences and the views that concern us. This example also bears on the relation between the Ramsey sentence of a theory and its "Craig transcriptions." I first learned of it from a paper of van Benthem (1978) who credits Swijtink (1976) with the idea. (A related, but more technical, example occurs in Ketland [2004].) I was recently reminded of Swijtink's example when Hilary Putnam shared with me his independent discovery of an almost identical construction.[2] The discussion which follows is the result of reflection on this example and conversations and correspondence with Putnam and others[3] on its possible applications and general interest.

We will restrict our attention to first-order theories with identity whose nonlogical vocabulary is partitioned into "observation" and "theoretical" terms. The language of the theory which is of particular interest to us has just one observation term and one theoretical term; the former is the binary relation symbol R and the latter is the unary predicate symbol A. Since the example is abstract and, at this stage, is intended to serve only logical purposes, we need not go into how the distinction in vocabulary is drawn, how it is justified, or whether it can be justified. We will consider these questions later.

The first-order languages with identity that are defined over the non-logical vocabularies {R} and {A} will be denoted by L_o and L_t, respectively.

---

[2] At the conference "Analysis and Interpretation in the Exact Sciences," held at the University of Western Ontario, May 2–4, 2008, and organized by Robert DiSalle.
[3] Especially Michael Friedman and Anil Gupta.

$L_o \cup L_t$ denotes the language generated by {A, R}. (Notice therefore that L_o $\cup$ L_t includes formulas that contain *both* observation and theoretical vocabulary items, as well as formulas of L_o and L_t.) I will use the terms L_o, L_t and $L_o \cup L_t$ to mean the particular languages I have just characterized, but I will also use these terms more generally for arbitrary first-order languages generated from observation, theoretical, and observation *and* theoretical vocabularies, respectively. The context should always make clear which is intended. Similarly, I will sometimes use T for the example I am about to describe, and sometimes for an arbitrary theory.

Let φ be an L_o-sentence which axiomatizes the complete theory of the natural numbers with <. Then the *toy theory* T (see van Benthem 1978, p. 324) is the theory in $L_o \cup L_t$ consisting of φ together with sentences expressing the fact that A is nonempty, that the first element (first with respect to <, of course) is not in A, and that the complement of A is closed under immediate successors. (This last condition means that if *x* belongs to the complement of A, then for any *y* such that *x*R*y*, if there is no *z* such that *x*R*z* and *z*R*y*, then *y* belongs to the complement of A.)

The L_o-consequences of T are evidently true in the standard model ⟨N, <⟩ of the natural numbers under <. But T holds only in nonstandard models, i.e. in models with infinitely large or nonstandard natural numbers.[4] Essentially the same is true of T's Ramsey sentence, $\exists X T(R, X)$, even though it is strictly weaker than T. This is because the complement of A must contain the first element of the <-sequence and must be closed under successors. So consider the standard second-order model ⟨N, $\wp(N_i)_{i \in I}$, <⟩, where for each *i*, $\wp(N_i)$ is the power set of the *i*-th Cartesian power of N. Such a second-order model is said to be *full* because the domain over which the second-order variables range is the true power set of the domain of individuals: in the present case, the power set of the set of natural numbers. For the Ramsey sentence of T to hold in ⟨N, $\wp(N_i)_{i \in I}$, <⟩, there must be a set in $\wp(N)$ which contains zero, is closed under successors, and does not contain every element of N. But this is impossible when ⟨N, <⟩ is standard.

Any theory in $L_o \cup L_t$ whose L_o-consequences are recursively enumerable has a *Craig transcription*, i.e. a recursive axiomatization of its L_o-consequences by sentences of L_o. As Putnam has noted, our T provides a particularly simple demonstration of the fact, first proved by Craig,[5] that

---

[4] The elementary facts about nonstandard models which I will assume can all be found in Enderton (1972, Chapter 3, sects. 1 and 2), and in greater detail in Robinson (1966, Chapter 2; and Chapter 3, sects. 1 and 2).

[5] Craig's observation and its proof were described to me by Hilary Putnam in conversation. Craig never published his observation or its proof.

although a theory and its Ramsey sentence – like a theory and any of its Craig transcriptions – have the same L_o-consequences, the Ramsey sentence of T can fail in a model in which a Craig transcription of T holds. There is therefore a sense in which T's Ramsey sentence tells us more than what is given by its observational consequences. This is because the Ramsey sentence distinguishes among models for the language L_o those that can be *expanded* to a model for the language $L_o \cup L_t$ to a model of the theory from those that cannot. (Here and elsewhere it is important to bear in mind that expansions and reductions of models differ from extensions and substructures since they do not change the domain of a model, but merely interpret or fail to interpret certain nonlogical constants.) A Craig transcription makes no such distinction, since it holds in all models of the L_o-consequences, including any that cannot be expanded to a model of the theory. We will return to this point.

Ramsey's own interest in Ramsey sentences did not address model-theoretic considerations of the sort just mentioned. For him the notion simply gave precise expression to the fact that the derivation of propositions belonging to what he called the "primary system" (the L_o-sentences in our terminology) does not depend on our assigning any meaning to the "secondary" or theoretical terms beyond the isolation of their logical category. As Ramsey wrote,

> We can say, therefore, that the incompleteness [that results when, in the secondary propositions, nonlogical constants are replaced by variables] affects our disputes but not our reasoning.[6]

Ramsey's point is that our disputes, in so far as they involve questions of truth and mutual compatibility, depend on the completeness of our propositions, and completeness is precisely what is set to one side when we pass from secondary propositions to the propositional functions which replace them under ramsification. But our reasoning with secondary propositions does not require their completeness. By 'our reasoning' Ramsey clearly intended to include both the derivation of consequences of the theory – especially the derivation of L_o-consequences – and the use of the theory to reason hypothetically to conclusions which, although not consequences of the theory, are conclusions that would not be forthcoming without its help. So far as our reasoning with the theory in any of these senses is concerned, the secondary vocabulary is entirely eliminable in favor of variables,

---

[6] Ramsey (1929, p. 232), hereafter referred to by its title, "Theories." Page numbers refer to the publication of "Theories" in Braithwaite (1931); to find the corresponding page number for the paper's reprinting in Mellor (1990), simply subtract 100.

provided evident consistency conditions are respected in their use. Moreover, the eliminability of the secondary vocabulary in the reconstruction of our reasoning does not require its *reduction by explicit definitions* to the primary vocabulary.

In an important early discussion of Craig transcriptions, Hempel observed that while a Craig transcription of a theory shares its L_o-consequences, it does not preserve those connections between observable properties and relations whose mediation by their connection with theoretical properties and relations is expressed by the theory.[7] Hempel went on to argue that although such connections are not necessarily reflected in the L_o-consequences of the theory, they can sometimes be exploited by inductive arguments that would be missed were it not for the connections between observable and theoretical properties that the theory expresses. These connections are lost when a theory is replaced by one of its Craig transcriptions, and so also, therefore, are the inductive arguments the connections enable.

Although Hempel emphasized the contrast between a theory and its Craig transcriptions, he did not observe that precisely the same contrast holds between a Craig transcription of a theory and its Ramsey sentence. One could easily be misled to conclude from Hempel's discussion that the source of the difficulty with Craig transcriptions is their elimination of theoretical terms. But the relative weakness of a Craig transcription is not due to its elimination of theoretical terms. Ramsey sentences also eliminate theoretical terms. Unlike the Ramsey sentence of a theory, a Craig transcription eliminates theoretical terms without providing analogues of the sentences which contain them.[8] Even though the vocabulary of the Ramsey sentence is wholly observational, it expresses the connections between observable properties that are mediated by theoretical properties and that are essential to the arguments Hempel described. Ramsey's achievement was therefore importantly different from Craig's. Ramsey sought to recover all secondary system-dependent reasoning to conclusions involving the primary vocabulary. Although such arguments occur within a framework that contains secondary vocabulary, Ramsey recognized that the reconstruction of our reasoning in this theoretical framework requires only the preservation of the logical category of the terms belonging to that

---

[7] See Hempel (1963, pp. 699–700) and Hempel (1958/1965, pp. 214–16).

[8] To see this, it is necessary to know how Craig established his observation about the recursive axiomatizability of the L_o-consequences. The proof uses a recursive enumeration of the L_o-consequences to define a recursive axiomatization by letting the first axiom be the first L_o-consequence in the enumeration and, in general, letting the $(n + 1)$-th axiom be the conjunction of the first n L_o-consequences of the enumeration. See Putnam (1965) for an exposition and references.

vocabulary. The terms themselves are entirely dispensable. It is in this sense that Ramsey showed the superfluousness of the secondary vocabulary.

7.3 RAMSEY SENTENCES AND THE APPLICATIONS OF A THEORY

As we noted earlier, the discovery of the notion of a Craig transcription of a theory raised questions about Ramsey sentences that go beyond their ability to capture our reasoning with the theory. Following van Benthem, call the *applications* of a theory T those models for the observation language L_o of T that can be expanded to models of T. The reason why Ramsey sentences suggest themselves as a natural tool for investigating applications in this sense is that it may happen that only infinite classes of L_o-sentences are capable of defining the class of models of interest. If T has only infinite models and is recursively axiomatized, then by allowing finitely many additional predicates, it is possible to finitely axiomatize the L_o-sentences of first order that characterize the applications of T.[9] Extending the logic to one of second order by existentially generalizing away the new predicates, we remain within a language whose nonlogical constants are observational. But we also have that the L_o-consequences of T are finitely axiomatized by a sentence that is *in the same form as* a Ramsey sentence.

Van Benthem (1978, pp. 328–9) gives an illustration of the Craig–Vaught results which we will find useful when we come to consider Ramsey's philosophical views. The complete theory T' of ⟨N, o, S⟩, with S the successor relation, is recursively axiomatized, but not finitely axiomatizable. In the standard recursive axiomatization, for every n there is an axiom which prohibits loops of successor-related elements of length n. Suppose we introduce a binary predicate R (whose intended interpretation is <) and the axiom

$$(^*)\forall x\forall y(\text{if } xSy, \text{ then } xRy)$$

stating that if x is succeeded by y then $x < y$. In the present context, R counts as "theoretical" and o and S as "observational." (This division into observation and theoretical vocabulary is completely arbitrary, but bear in mind that at this stage we are considering only abstract and logical aspects of ramsification, and that we have yet to deal with the question of the epistemological basis for the distinction between observation and

---

[9] These results are due to Craig and Vaught (1958). They occur in van Benthem (1978) as Theorems 3.10 and 3.12.

theoretical vocabulary.) If, in addition to the axiom (*) relating < to successor, we add axioms stating obvious properties of <, the result is a finite axiomatization T″ of T′ in which (*) replaces the infinitely many axioms prohibiting loops. R is not definable in terms of o and S (recall that T′ and T″ are first-order), but if we ramsify T″, we are back to a theory whose nonlogical vocabulary is wholly observational.

Craig and Vaught's observations can be usefully compared with the use of Ramsey sentences in the sense of Ramsey. Suppose we are given a theory T that finitely axiomatizes its L_o-consequences. (This assumption is not essential to the comparison, but making it avoids raising extraneous issues.) The Ramsey sentence of T and the Ramsey sentence of a Craig–Vaught theory that finitely axiomatizes T's L_o-consequences by the addition of finitely many new parameters are both second-order finite axiomatizations of the L_o-consequences of the original theory in the vocabulary of L_o. But each of the models of the L_o-consequences of the Ramsey sentence of the Craig–Vaught theory can be expanded to a model of the theory. By contrast, this may or may not be true of the models of the L_o-consequences of the Ramsey sentence of the original theory T. So there is an interesting class of theories, namely those that are recursively axiomatized and have only infinite models, with the property that all of their applications are characterized by "their" Ramsey sentences, where the use of scare quotes around 'their' is meant to signal the fact that the original theory may have been modified in recovering a finite axiomatization of its L_o-consequences. This modified theory has the property that every model of the original theory's L_o-consequences can be expanded to a model of it, but this is something that need not be true of the original theory or its Ramsey sentence.

The philosophical interest of these results regarding the finite axiomatization of the L_o-consequences is admittedly modest, since it is reasonable to assume that the original notion of a Ramsey sentence, even requiring as it does a finitely axiomatized theory, is applicable to most cases that are likely to be of interest to a general investigation of the nature of theories. But they at least show that under certain natural assumptions regarding the intended scope of one's analysis, the sentences that characterize a theory's applications can be finitely axiomatized. And they show that under these conditions the new parameters used in the finite axiomatization of the L_o-consequences are eliminable by ramsification, so that the axiomatization can be replaced by a sentence which is in the form of a Ramsey sentence. However, the models of this theory might not coincide with the models of the original theory.

7.4 EXTENSIONS AND EXPANSIONS OF MODELS AND RUSSELL'S STRUCTURALISM

Let us turn to a closer consideration of a nonstandard model of the toy theory T. The binary predicate R is interpreted by < on pairs of standard numbers, and the predicate A by a set of nonstandard numbers. So the domain of a nonstandard model of T consists of a standard and a nonstandard part. Call the standard part N_o, and the nonstandard part N_t. The domain of the model is then $N^* = N_o \cup N_t$. The interpretation of A is wholly included in the nonstandard part of the domain, and R is interpreted by a relation $<^* \subseteq N^* \times N^*$ which is such that its standard part (its restriction to $(N_o \times N_o)$) is <. A "general" second-order structure in which the Ramsey sentence of T holds has the form $\langle N^*, S(N^*_i)_{i \in N}, <^* \rangle$, where for each i, $S(N^*_i)_{i \in N}$ is a (possibly proper) subset of the power set of the i-th Cartesian power of N^*, and there is an assignment of a proper subset of N_t to the variable X in T(X, R) under which $\exists X T(X, R)$ comes out true in $\langle N^*, S(N^*_i)_{i \in N}, <^* \rangle$.

The relation $<^*$ which orders the nonstandard numbers also satisfies many of the conditions imposed on the standard <-relation which holds among standard numbers, but the order-type of the nonstandard model differs from that of the standard numbers. It might be thought that because of this difference, the Ramsey sentence of a theory such as T, requiring as it does a nonstandard model for its truth, involves a change in the intended meaning of the observation vocabulary. But this conclusion would be too quick. Among other things it overlooks the reliance of at least one of the philosophical theories that interests us on an assumption that is of fundamental importance to its conception of our epistemic predicament. To see how this happens, we must turn to Russell's articulation of his theory of knowledge in *The Analysis of Matter*.

Russell's adherence to a particular form of empiricism justifies presenting his theory of knowledge as one that depends on a partition of descriptive vocabulary into observation and theoretical terms. Given the nature of his theories of propositional understanding and perception, the basis for this distinction rests on the intended reference of the observation and theoretical vocabulary items. In greater detail, from his theory of meaning Russell derived the "fundamental principle for the analysis of propositions," according to which any proposition we can understand must be composed wholly of "constituents" with which we are acquainted.[10] By his theory of perception,

---

[10] 'Proposition' should be understood in the sense of a Russellian proposition. This is a structured entity whose constituents are individuals and the properties and relations which qualify them. The "fundamental principle" is formulated particularly clearly in Russell (1912, p. 58).

the realm of "events" with which we are acquainted is restricted to quasi-mental events – what Russell calls "percepts." We are capable of knowing the properties and relations of events with which we are acquainted, and so they too may be included among items of acquaintance. But our knowledge of properties and relations of events with which we can have no acquaintance requires a different account. This is what the theory of *The Analysis of Matter* promises to provide. Taken in combination with his theory of descriptive knowledge, it preserves the primacy of acquaintance while permitting the possibility of knowledge of things that are incapable of being constituents of the propositions we understand.

To illustrate Russell's view using our toy theory T, let the standard natural numbers in N_o correspond to events with which we are acquainted, and let the nonstandard numbers in N_t correspond to events which transcend our acquaintance. According to Russell, we can have full knowledge of any relation which holds among events with which we are acquainted, but because we are unacquainted with theoretical events, we cannot claim full knowledge of the relation which orders them, but can say only that it is a binary relation with certain of the logically or mathematically expressible properties the <-relation exhibits in the observable part of the domain: understood as a relation involving theoretical events, $<^*$ is known only as a relation that shares certain of the formal properties of the <-relation among percepts. In agreement with common sense, Russell holds that there are theoretical entities and relations among them, and that this assumption is explanatory of our experience in a way in which an account which confines itself wholly to events with which we are acquainted is not; where he departs from mere common sense is in his claim that all we can know of relations involving theoretical events is their abstract structure.

Whatever the truth about Russell's theories of perception and propositional understanding, the distinction in vocabulary which they mandate is not arbitrary; nor is it undermined by the observation that it fails to match our conventional understanding of 'observational' and 'theoretical.' Russell's project is to recover a facsimile of what passes for our knowledge of the world within the framework of epistemic constraints imposed by his theories of perception and propositional understanding. The criteria for the success of the project allow for the possibility that when reconstructed, the character of our recovered knowledge may differ considerably from what, pre-theoretically, we understood it to consist in. Nevertheless, there must be *some* limit to the extent to which a reconstruction of the character of our knowledge is allowed to differ from our pre-analytic conception, and this brings us to Newman's well-known criticism of Russell's theory in Newman

(1928). As we will see, its susceptibility to this criticism undermines the epistemological significance Russell's project attaches to the distinction between observation and theoretical vocabulary.

Newman's criticism can be expressed very simply within an elementary model-theoretic framework. The formulation depends on a "folklore" result of mathematical logic which appears in van Benthem (1978) as Lemma 3.2:

> For any L_o-structure M, if M is a model of the L_o-consequences of T, then there exists an $L_o \cup L_t$-structure N such that N is a model of T and the reduction of N to L_o is an elementary extension of M.

Let M be the L_o-structure whose domain is the domain of observable events on Russell's theory. We must show that the domain M of M has an extension M* which is the domain of a model of T, where M* is the set of observable and theoretical events. But for M* to form the domain of a model of T – and hence for T to be *true* – it suffices that there should be a one–one mapping from the domain N of the model N (which exists by the lemma) to M*. This consequence is justified by the fact that there is nothing in Russell's theory to prevent using an arbitrary one–one mapping from the "theoretical part" of N to the theoretical events of M* to *define* the theoretical relations of the model M* as the images of the theoretical relations of N – relations which we know from the lemma to satisfy the theory. Thus we have *Newman's objection* to Russell's theory:

> The claim that our knowledge of the theoretical part of the world is purely structural in the sense of Russell's theory – i.e. in the sense that we can say of the relations which hold there that they share certain formal properties with known relations but are otherwise ignorant of what these relations are – is true as a matter of logic and an assumption about the cardinality of the world of observable and unobservable events. Hence, if Russell's theory is intended to capture the character of what we normally take to pass for theoretical knowledge, it has failed.

Notice that this formulation of Newman's objection depends only on the folklore result recorded in van Benthem's lemma and an assumption about the cardinality of the domain of all events, both theoretical and observable. In particular it does not depend on anything so recherché as the Löwenheim–Skolem theorem, let alone any of its various generalizations. Cardinality enters the argument only as an empirical assumption that may be false: in fact, as the only empirical assumption that the truth of claims about the theoretical part of M* depends on.[11]

---

[11] The idea of presenting Newman's objection model-theoretically was first suggested in Demopoulos and Friedman (1985, sect. 3), where its connection with the "model-theoretic argument" of Putnam (1977) is also discussed. The model-theoretic formulation of Newman's objection is elaborated and applied to van Fraassen's constructive empiricism in Demopoulos (2003a, sect. 5), reprinted as

Let us consider again the L_o-consequences of T and the *standard* model $\langle N, < \rangle$: while this structure cannot be *expanded* to a model of T, the lemma tells us that it has an elementary *extension* to a model of T. As we have presented it, Newman's criticism of Russell's theory depends on a lemma about extensions (rather than expansions) and an assumption about the cardinality of the domain of theoretical events. The appeal to an extension is internally justified by Russell's own theory, since the realist component of that theory consists in the assertion that there is an extension of the domain of events that are known by acquaintance to one that includes theoretical events; the structuralist component asserts, on the basis of the satisfaction of this realist assertion, a particular proposal regarding the nature of our knowledge of the theoretical events which belong to this extension. The intention of the realist component is to strengthen the theory consisting of just the L_o-consequences of T by the addition of assertions about things falling outside our acquaintance. But the structuralist component's account of the nature of the knowledge expressed by such assertions is actually in tension with the realist component, since it has the consequence that statements about the theoretical part of the domain are true as a matter of logic and an assumption about cardinality, a consequence that vastly understates the intended epistemic significance of theoretical statements and threatens the robustness of Russell's claim to realism about the theoretical part of the domain.

The technical distinction between extensions and substructures versus expansions and reductions figures in a recent discussion of contemporary formulations of Russell's structural realism. Ketland (2004) introduces a distinction between two notions of empirical adequacy. Ketland calls a theory *weakly empirically adequate* if all of its L_o-consequences are true. He calls a theory *empirically adequate* if it has an "empirically correct full model," a condition which, in an arithmetical context like the one we have been assuming in our examples, means that a theory is empirically adequate only if it has a model whose *reduction* to L_o is isomorphic to the standard

---

Chapter 6 of this volume. Van Fraassen (2008, pp. 230–2) argues that Putnam's model-theoretic argument establishes the apparent paradox that almost any consistent set of sentences is true. It is important to notice that van Fraassen's use of Putnam's argument is not restricted to theoretical sentences. Nor is the difficulty it is held to raise specific to a particular philosophical view. Rather, the fact that for almost any consistent set of sentences it is possible to construct an interpretation which makes them true is presented as a challenge to our pre-analytic intuition that our sentences have a definite content. I believe that van Fraassen's use of Putnam's argument cannot be justified. But the point I wish to emphasize here is that despite some overlap in the formal considerations they adduce, van Fraassen's analysis has almost nothing in common with the analysis presented in the text. *Added in 2012*: For an extended discussion of van Fraassen (2008) on these matters see the appendix to Chapter 4, above.

model. Thus, the observation to which T led us, regarding extensions and expansions of models of the L_o-consequences of T to nonstandard models of arithmetic, is essentially what underlies Ketland's distinction between the notions of weak empirical adequacy and empirical adequacy: our theory T is *weakly* empirically adequate, but not empirically adequate, because none of its models is standard, and hence none of them has a reduction to L_o that is isomorphic to the standard model.

Ketland shows that if we confine our attention to theories that are empirically adequate, then the Ramsey sentence of any such theory is true in any full second-order model of the theory's observational consequences provided that the domain of the model has the right cardinality, i.e. provided there is a one–one correspondence between the theoretical part of the domain of the model of the theory's L_o-consequences and the "theoretical part" of the domain of the model which we know to exist on the basis of a purely logical argument. To turn the model of the L_o-consequences into a model of the Ramsey sentence of the theory, one uses the construction of theoretical relations in terms of the one–one correspondence that we sketched earlier.[12] Ketland emphasizes that the same claim cannot be made about the Ramsey sentence of a weakly empirically adequate theory, as the example of T shows: the Ramsey sentence of T fails in the standard second-order model $\langle N, \wp(N_i)_{i \in N}, < \rangle$. For this reason Ketland restricts his criticism of structural realism to its analysis of empirically correct theories, where it can be shown that this analysis has the consequence that any model of the L_o-consequences of such a theory can be *expanded* to a model of the Ramsey sentence of the theory provided only that its domain is large enough.

It is important to recognize that Ketland's restriction to empirically correct theories is clearly motivated only in the case of arithmetical examples like T where it effectively excludes theories with nonstandard models. But the considerations that argue in favor of standard interpretations in the context

---

[12] In his valuable review of responses to Newman, Ainsworth (2009, p. 144) writes that:

> [Demopoulos and Friedman] claim (although without substantial argument) that if 'our theory is consistent, and if all its purely observational consequences are true, then the truth of the Ramsey sentence *follows* as a theorem of set theory or second-order logic, provided our initial domain has the right cardinality' (Demopoulos and Friedman 1985, p. 635, original emphasis). Although Demopoulos and Friedman do not really back up this claim, Ketland (2004) does provide a strong argument for (a slight variant of) the claim.

> But the core of the argument for this claim *is* given in Demopoulos and Friedman (1985, pp. 633–4). This is where Putnam's model-theoretic argument is discussed and where it is noted that Newman's observation depends on nothing more than what Ketland calls the "push forward" construction used by Putnam.

of the theory of models of arithmetic, such as the "naturalness" of the standard model, do not transfer to theory of knowledge or theory of theories. For example, Ketland argues for the correctness of his condition of empirical adequacy by citing van Fraassen's idea that:

a theory is empirically adequate exactly if what it says about the observable things and events in this world, is true – exactly if it "saves the phenomena." A little more precisely: such a theory has at least one model that all the phenomena fit inside. (van Fraassen 1980, p. 12, quoted by Ketland 2004, p. 296)

However, the pre-analytic intuition that this passage expresses – specifically, having at least one model that all the phenomena fit inside – is neutral between an *expansion* of a model of the L_0-consequences of T to a model of T and an *elementary extension* of such a model to a model of T. As we saw, the latter notion applies even when the model of the L_0-consequences is standard and the model of T is necessarily nonstandard. So although Ketland's distinction between empirical adequacy and weak empirical adequacy is technically interesting, it is irrelevant to the critical discussion of the views of Russell or the structural realists who follow him. For as we saw earlier, even if $\langle N, < \rangle$ fails to have an expansion to a model of T, it has an elementary extension to a model of T, and this elementary extension can be taken as the basis for a second-order model of the Ramsey sentence of T. The appeal to such an extension merely makes explicit the assumption that the domain of the model of the observational consequences has been enlarged to include theoretical entities. But this can hardly be an objection, since the realist component of the structuralist view already endorses such an extension.

7.5 RAMSEY ON THEORIES

"Theories" is a detailed working out of an artificial example of a theory which describes the temporal series of perceptions of color one would have if one occupied one of three possible locations in a very simple world. The point of the paper is far from transparent, and the philosophical considerations motivating the discussion are not nearly as explicit as one would like. The notion of a Ramsey sentence is introduced only toward the end of the paper and without the emphasis on its possible significance that has accompanied its introduction by subsequent advocates of the notion's utility. Any claim to have extracted a philosophical theory from Ramsey's paper that captures his intentions would be tendentious. The following account is strongly suggested by many of Ramsey's explicit statements, but I do not claim that it is forced by them. However, I do claim that an expository virtue of my proposed reconstruction of the views of Russell, Ramsey, and Carnap is that the philosophical theory it imputes to Ramsey presents an interesting contrast with Russell's, and this contrast provides a natural context for Carnap's proposals for the interpretation of Ramsey sentences.

In a paper read in 1925 to a Cambridge discussion society Ramsey remarked,

I don't really believe in astronomy, except as a complicated description of part of the course of human and possibly animal sensation.[13]

And "Theories" opens with the sentence,

Let us try to describe a theory simply as a language for discussing the facts the theory is said to explain. (p. 212)

If we extrapolate from these remarks to a view of theories in general, then capturing the possible applications of a theory by extending the primary vocabulary by finitely many additional parameters would seem to come close to capturing the sentiment they express. Ramsey's view of the role of theories and the interest he attached to ramsification would then come down to the idea that a potentially infinite description in terms of primary propositions is rendered manageable – is finitely axiomatized – by the addition of finitely many new "theoretical" parameters. On such a view, the role of theory is to provide organizing parameters, and when the principles such parameters satisfy are ramsified, we are left with a proposition all of whose nonlogical vocabulary is primary. The "content" of the theory would then be properly reflected by the primary system and the nonlogical vocabulary it requires to describe the course of human and possibly animal sensation.

Our discussion of T′ and T″ and of the standard and nonstandard models of the L_o-consequences of T can further clarify, at least at an abstract level, Ramsey's conception of the role of the secondary system and its theoretical vocabulary. By contrast with Russell, for Ramsey it is the addition of new parameters that do *not* require for their interpretation an extension of the observable part of the domain that is particularly well suited to illustrating the role of the secondary system. For according to this view, the purpose of the secondary system and the theoretical vocabulary it introduces is to organize our knowledge of the things with which we are

---

[13] Ramsey (1925a, p. 291) in Braithwaite (1931) and Mellor (1990, p. 249).

acquainted. But there is no presumption that in doing so we are partially disclosing the character of entities that are hidden from our acquaintance. For Ramsey the secondary system introduces new parameters that are defined on the domain of any application of the theory, but by including these parameters we do not necessarily enlarge the domain of any such application. On this interpretation, Ramsey's conception of the secondary vocabulary is perfectly illustrated by our earlier discussion (in Section 7.4) of the role of the theoretical predicate R in recovering T′ from T″.

Remember that by limiting models of T's L_o-consequences to expansions of $\langle N, < \rangle$ to models for $L_o \cup L_t$ that are models of T we exclude nonstandard interpretations of L_o. When T has only nonstandard models, T can be false when its L_o-consequences are true, if by 'true' we mean *true in the standard model*. As we saw, T's Ramsey sentence also cannot be modeled by a second-order model $\langle N, \wp(N_i)_{i \in N}, < \rangle$ which preserves the standard interpretation of L_o, so it too fails to be true. It is therefore at least consistent with Ramsey's remarks to suppose that he would respond to cases of this kind with an appeal to what I will call *Ramsey's principle*:

If the Ramsey sentence of a theory can only be modeled by an *extension* of the domain of a model of its L_o-consequences, then this is grounds for rejecting the theory; such a theory requires an unjustifiable modification of the structure of the models of the L_o-consequences and a larger domain than what is strictly demanded by the truth of its L_o-consequences.

The implicit philosophical theory of Ramsey's principle is very close to the reductionist program that we associate with Russell's phenomenalist period: roughly, the period 1914–19, and principally represented by *Our Knowledge of the External World*.[14] Russell conceived the central task of the program he then advocated to require showing that "physical objects are logical constructions out of sense data." Ramsey noticed that extending Russell's program to include an account of the nature of theories would require the explicit definition of the terms belonging to the theoretical vocabulary in terms of the primary vocabulary. So understood, Ramsey's paper is a natural development of Russell's phenomenalism since it shows how to achieve the *effect* of the program of logical construction in the context of a theory of theories – namely, the elimination of the theoretical vocabulary – without having to enter into the tendentious issue of what to count as an acceptable defining formula, and without having to construct suitable definitions, however the

---

[14] Russell (1914a). The end of Russell's commitment to this program is clearly marked in Russell (1919), especially pp. 61–2 of that work, quoted above in Section 5.1.

notion of definition is understood. So represented, Ramsey's contribution to Russell's phenomenalist program is reminiscent of his proposal that to solve the difficulties the Axiom of Reducibility poses for Russell's logicism we should abandon the ramified theory of types in favor of the simple theory. Both proposals involve a reduction in the technical requirements that Russell supposed were demanded by the successful implementation of a philosophical program, and both proposals focus on Russell's emphasis on the necessity of various types of definitions: explicit definitions in the case of the extension of his phenomenalism to a theory of theories, and definitions which respect the intricate system of orders of the ramified theory in the case of his logicism.

The above account of Ramsey's philosophical view differs in some respects from the interpretation advanced by Richard Braithwaite, Ramsey's friend and the original editor of his papers.[15] The first difference is largely one of emphasis and concerns Ramsey's relation to Russell's logical constructionist program. For Braithwaite, the principal lesson of "Theories" is to have revealed the limitations of Russell's program of logical construction.[16] Theoretical concepts are explicitly definable in terms of the concepts of the primary system only if the secondary system cannot be properly adapted to apply to new situations, and since the theories of greatest interest *are* capable of being adapted to new situations, Russell's logical constructions are of little value for philosophy of science:

> It is only in theories which are not intended to have any function except that of systematizing empirical generalizations already known that the theoretical terms can harmlessly be explicitly defined. A theory which it is hoped may be expanded in the future to explain more generalizations than it was originally designed to explain must allow more freedom to its theoretical terms than would be given them were they to be logical constructions out of observable entities. A scientific theory which, like all good scientific theories, is capable of growth must be more than an alternative way of describing the generalizations upon which it is based, which is all it would be if its theoretical terms were limited by being explicitly defined.[17]

Suppose we grant that this objection can be extracted from Ramsey's paper. And suppose we also grant Braithwaite's claim that the explicit definability of the secondary vocabulary in terms of the primary vocabulary is

---

[15] Braithwaite's own views (some of which we will discuss below) were significantly influenced by Ramsey. As he remarks in the preface to *Scientific Explanation*:

My thinking about the role of theoretical concepts and of the philosophy of science in general was re-oriented when I had the melancholy privilege of editing F. P. Ramsey's brilliantly suggestive papers after his death in 1930. (Braithwaite 1953, pp. ix–x)

[16] Braithwaite (1953, p. 52) asserts that Ramsey's aim was the refutation of Russell.

[17] Braithwaite (1953, p. 76). The same objection to logical constructions occurs in Beck (1950).

incompatible with the growth of science. It is important to recognize that neither concession affects our claim that Ramsey's paper can be appealed to in support of an extension of Russell's phenomenalism to a theory of theories. Ramsey's explicit remarks about Russell and others – the relevant passage from "Theories" (p. 220) cites Carnap, Nicod, and Whitehead as well – take issue only with the thesis that to justify the *elimination* of theoretical terms it is necessary to find explicit definitions for them in terms of the primary vocabulary. There is of course a difference in the technical methods Russell proposed and those that Ramsey thinks are sufficient for the philosophical goals of phenomenalism: against Russell, Ramsey argues that to eliminate theoretical vocabulary it is unnecessary to settle on what should count as the correct notion of definition or to address the thorny technical problem of providing suitable definitions in terms of the primary vocabulary. But the phenomenalist contention that the secondary vocabulary is eliminable is one that Ramsey's analysis was intended to *support*.

The second difference is not one of emphasis. It concerns the issues raised by Ramsey's principle and the legitimacy of *extensions* of the domains of the applications of a theory. Here it is important to recognize that Braithwaite's rejection of the goal of systematizing only known generalizations expressed in the primary vocabulary leaves entirely open the possibility that, on Ramsey's view, the goal of theory is the systematization of truths, among them truths that would be unknown were it not for the invocation of theory, but all of them truths which are expressed in just the primary vocabulary. It follows that one might consistently hold that the function of a theory is not exhausted by the goal of systematizing *known generalizations*, and yet reject as illegitimate certain kinds of theoretical extensions of the domains of a theory's applications, arguing that even if the aim of theory is not the recovery of known generalizations expressed in the primary vocabulary, its aim is the systematization of *truths* expressed in that vocabulary. His different perception of the role of extensions in the pursuit of this aim is what I earlier argued separates Ramsey's views from the realism Russell embraced in *The Analysis of Matter*.

Braithwaite comes closest to addressing the issue of extensions when he considers what he calls 'the question of the "reality" of the entities our theoretical terms denote' (Braithwaite 1953, p. 79). Braithwaite begins his discussion by endorsing an answer to this question which he says is in essence the one given by Ramsey:

[T]he status of a theoretical concept (e.g. an electron) is given by the following proposition which specifies the status of an electron within the deductive system of contemporary physics: There is a property E (called 'being an electron') which is such that certain higher-level propositions about this property E are true, and from these higher-level propositions there follow certain lowest-level propositions which are empirically testable. Nothing is asserted about the nature of this property E; all that is asserted is that the property E exists, i.e. that there are instances of E, namely electrons. (*Ibid.*)

But what are the instances of E to which we are committed but of whose nature nothing is asserted? Braithwaite proposes two answers to this question, one that he finds in Ramsey, and another of his own. He attributes to Ramsey the view that to assert that there are electrons is merely to assert the truth of the theory in which electrons occur. But if this is Ramsey's answer, it is unhelpful, since our question is equivalent to asking how the truth of the theory is to be understood – equivalent, that is, to asking whether or not the truth of the theory requires extensions of the domains of its applications. Braithwaite effectively concedes this, and this concession leads him to propose another way of addressing the question of the "reality" of electrons:

This other way of answering the question . . . is to side-step [it] and to make no remark whatever about a theoretical concept itself as opposed to a theoretical term. Instead of answering the question directly, an answer is given which refers not to the concept of electrons . . ., but to the word 'electron' . . . and to the use of [this symbol] in the [theory]. (*Ibid.*, p. 80)

The difficulty with this suggestion is that one obvious use of the term 'electron' is to assert the extension of the domain of an application of a theory to which theoretical postulates containing 'electron' have been added. Directing us to the use of the theoretical *term* rather than to the theoretical *concept* does nothing to side-step the question of the reality of electrons.

7.6 CARNAP'S SYNTHESIS

Earlier I claimed that Carnap's understanding of Ramsey sentences in his mature reconstruction of the language of science yields a kind of resolution of the realist and anti-realist differences between Russell's and Ramsey's respective accounts of theoretical knowledge. We are now in a position to see how this happens. Let us begin with Carnap's explication of the factual content of a theory.

Carnap may have shared the concept-empiricist inclinations of Russell and Ramsey, but for him the distinction between the observational and theoretical levels was not tied to theory of perception as it clearly was for Russell, and very likely was for Ramsey. For Carnap the role of the observation vocabulary is to contribute to the description of the evidential basis of our theoretical claims. As such it has neither the mentalist nor phenomenalist overtones that it had for Russell and Ramsey. This is an important difference, but it is not one that affects a basic point of agreement, namely that for Carnap, just as much as for Russell and Ramsey, the observation vocabulary is fully understood in a way that the theoretical vocabulary is not. This is why Carnap holds that there is a problem of "theoretical analyticity" that does not arise for the observation language: were the theoretical vocabulary as completely understood as the observation vocabulary, we would be in a position to judge of every sentence involving the theoretical vocabulary whether or not it is true in virtue of the meanings of its terms. According to Carnap, the fact that we are unable to make this kind of sharp distinction at the level of sentences containing theoretical terms shows such terms to be incompletely understood. Carnap's reconstruction of the language of science therefore has two tasks: to characterize the factual content of a theory and to sharply separate a theory's factual component from its analytic, nonfactual component. Our focus here is restricted to Carnap's reconstruction of the factual component of a theory, and his proposal that it is captured by the theory's Ramsey sentence.[18] For Carnap, the principal argument in favor of his reconstruction is that the Ramsey sentence of a theory has the same L_o-consequences as the theory itself.

In the lecture in which he first announced his Ramsey sentence reconstruction of the factual content of a theory,[19] Carnap asserted:

One of the most important characteristics of the T-terms, and therefore of all sentences containing T-terms – at least if they occur not in a vacuous way – is that their interpretation is not a complete one, because we cannot specify in an explicit way by just using observational terms what we mean by the 'electromagnetic field.' We can say: if there is a distribution of the electromagnetic field in such and such a way, then we will see a light-blue, and if so and so, then we will see or feel or hear this and that. But we cannot give a sufficient and necessary condition entirely in the observational language for there being an electromagnetic field, having such and such a distribution. Because, in addition to observational consequences, the content is too rich; it contains much more than we can exhaust as an observational consequence. (Psillos 2000, p. 159)

As I understand him, when Carnap says that the content of a theoretical construct like the electromagnetic field is too rich to be exhausted by its observational consequences, he is alluding to the fact that its content cannot

[18] For a discussion of Carnap's reconstruction of the analytic component, see Demopoulos (2007).
[19] Delivered at the 1959 American Philosophical Association meetings in Santa Barbara. The lecture has been transcribed and published as Psillos (2000).

be captured by a reconstruction which *merely* preserves the L_o-consequences of the theory to which it belongs. It will be recalled that this is precisely where a reconstruction of the factual content of a theory by its Ramsey sentence represents an advance over a reconstruction based on one of its Craig transcriptions. So on the matter of what is required to preserve our reasoning with the theory, Carnap is in agreement with Ramsey and Hempel. This agreement is evidently what led him to the Ramsey sentence and the thesis that theoretical terms are inessential to the representation of the factual content of the theory, even if they are not eliminable by the provision of a "sufficient and necessary condition" (explicit definition) expressed entirely in the vocabulary of the observation language.

But on the issue of whether theoretical claims must be understood as extending the domain of observable entities, or whether, with Ramsey, we should regard them as merely effecting a more tractable representation of our observations, Carnap advances a position that is neither Russell's nor Ramsey's. Even if Ramsey's observation about the role of the secondary vocabulary in theoretical reasoning is correct, by ignoring theories whose secondary vocabulary extends the domains of models of its observational consequences, Ramsey's account leaves out many theories of interest. Thus, contrary to Ramsey's phenomenalist inclinations, Carnap endorses without hesitation Hempel's remark in Hempel (1958/1965, sect. 9) that although the Ramsey sentence eliminates theoretical terms, it does not eliminate reference to theoretical entities. But contrary to Russell, Carnap proposes a deflationist interpretation of this concession. Thus he says of the Ramsey sentence that while he agrees with Hempel that it

does indeed refer to theoretical entities by the use of abstract variables . . . these entities are . . . purely logico-mathematical entities, e.g., natural numbers, classes of such, classes of classes, etc. Nevertheless, [the Ramsey sentence] is obviously a factual sentence. It says that the observable events in the world are such that there are numbers, classes of such, etc., which are correlated with the events in a prescribed way and which have among themselves certain relations; and this assertion is clearly a factual statement about the world. (Carnap 1963b, p. 963)

This reply expresses a view of theoretical properties and relations that stands in contrast with Russell's idea that the higher-order variables range over a domain composed of unobservable events and processes. Such an interpretation is reinforced by a subsequent discussion of the issue where Carnap explains that physicists may, if they so choose,

evade the question about [the existence of electrons] by stating that there are certain observable events, in bubble chambers and so on, that can be described by certain mathematical functions, within the framework of a certain theoretical system ... [T]o the extent that [the theoretical system] has been confirmed by tests, it is justifiable to say that there are instances of certain kinds of events that, in the theory, are called 'electrons.' (Carnap 1966 and 1995, pp. 254–5)

Hence on Carnap's view the extensions to which we are committed are to be found in the mathematical background of our linguistic framework. To avoid the "metaphysical" question of realism versus instrumentalism which divides Russell and Ramsey, Carnap embraces a view which is like Russell's in countenancing extensions of the domain to include more than just observable events, but is unlike Russell's in allowing for the possibility that theoretical extensions can be understood entirely in terms of the number- and set-theoretical apparatus of the representational framework within which the theory is formulated. To run Newman's objection against Carnap it is unnecessary to appeal to an empirical assumption about the cardinality of the domain of theoretical events, since for Carnap the "abstract" model **N** of the lemma that figures in that objection *already* suffices for the interpretation of the theory.

7.7 THE STRUCTURALIST THESIS

Let us pause and take stock of the conclusions we have reached. Although Russell never explicitly proposed a reconstruction of theories in terms of a language whose nonlogical vocabulary is partitioned into observation and theoretical terms, still less a reconstruction of the factual content of theories by their Ramsey sentences, his conception of theoretical knowledge included the fundamental assumption of any such reconstruction. Let us call this the *structuralist thesis*:

The theoretical component of what our theories express is wholly captured by statements which depend only on the logical category of their constituent concepts.

For Russell, the structuralist thesis has the consequence that our knowledge of events which transcend our acquaintance is knowledge only of their structure. For Ramsey it has the consequence that the secondary vocabulary is not essential to our reasoning to conclusions regarding the primary system – even if the secondary vocabulary is not eliminable by explicit definitions in terms of the primary vocabulary. For Carnap it has the consequence that it is possible to isolate the factual component of a theory by a sentence whose interpretation is neutral between the "metaphysical" alternatives of realism and anti-realism regarding theoretical entities.*

We have seen how Russell's view of theoretical knowledge is compromised by Newman's objection. And although Ramsey's views are not directly affected by Newman's observation in the way Russell's are, it is puzzling why anyone who accepts the structuralist thesis should oppose Russell's realism regarding the theoretical, given how weak Newman showed Russell's realism to have become when it is understood in the context of that thesis. Carnap's emendation succeeds in addressing Ramsey's omission of the existence assumptions that arise in important cases of fundamental physical theories, and it does so without succumbing to one or another of the "metaphysical" doctrines of realism or anti-realism. But this resolution of the incompleteness of Ramsey's account with Russell's realism comes at a price: on anyone's view, it must be conceded that any model of the L_0-consequences of a theory can be extended to a model of the theoretical claims of the theory. But since Carnap endorses the structuralist thesis, his understanding of the *truth* of theoretical claims is one according to which their truth is the same as their *satisfiability* in such a model.

Neither the structuralist thesis nor the difficulties it faces derives from the use of Ramsey sentences in the reconstruction of theoretical knowledge. It is rather that the structuralist thesis enables the claim that the factual content of a theory is captured by its Ramsey sentence. This was already clear to Braithwaite, whose *Scientific Explanation* contains the earliest discussion of the notion of a Ramsey sentence and its potential utility for the analysis of theories. The central idea of Braithwaite's own view is that a theory's purely theoretical propositions comprise an "abstract calculus," an idea that certainly commits him to the structuralist thesis. He comes close to recognizing a possible difficulty with the thesis in the section of his book entitled, "Are Campbellian hypotheses logically necessary?," where a Campbellian hypothesis is a sentence all of whose vocabulary is theoretical. The question is pressing for Braithwaite since such sentences may also occur in a "pure" deductive system or abstract calculus which is not integrated with sentences containing observation terms, and the consequences of such a pure calculus are "without question . . . logically necessary" (Braithwaite 1953, p. 101). The example Braithwaite describes is one where the theoretical sentences are consequences of the geometrical theorem that two circles cannot have three

---

* *Note added in 2012*: But see my discussion of Carnap's understanding of the Ramsey sentence of a theory in Section 3.4. Carnap's views on realism are critically evaluated in Chapter 4.

points in common. It might therefore be better characterized as a case in which Campbellian hypotheses are logical consequences of axioms that are rightly regarded as *analytic*, being in the present case axioms of *pure* geometry. In any case, Braithwaite goes on to argue, correctly in my view, that the possibility of such an interpretation of the purely theoretical sentences does not suffice to show them to be logically necessary or analytic, since they may also be interpreted so that the truth of the propositions they express is evidently contingent.

The situation Braithwaite has in mind can be more simply illustrated by a slight modification of our description of the observable part of the toy theory T. Suppose we restrict our attention to just the sentence φ which axiomatizes the complete theory of the relation < on the natural numbers, and suppose we modify the assumptions of our example by counting the relation symbol R as theoretical rather than observational. Then when φ is interpreted as the theory of the natural numbers under <, its consequences are "logically necessary" in Braithwaite's sense. However, we can also imagine an interpretation according to which φ describes an ω-sequence of theoretical entities ordered by a relation with the properties of <; in this case, both the fact that the objects are configured in this way and the fact that the relation that orders them has the properties of < are contingent matters of fact. So Braithwaite is entirely correct to maintain that his account of purely theoretical propositions does not make them logically necessary or analytic.

Braithwaite is entitled to this conclusion because the structuralist thesis refrains from specifying anything more than the logical category of a theoretical concept. It cannot be maintained against the thesis that it represents theoretical statements as true in virtue of the meanings of their constituent terms, since, aside from their logical category, *nothing is assumed about the meanings of their terms*. Braithwaite and other adherents of the structuralist thesis have focused on this feature of their view of theoretical terms and on the fact that the domain of any model of the theoretical claims of a theory is constrained by the requirement that it must satisfy the theory's observational consequences. Evidently not every model can fulfill this condition. But restricting our attention to those that do, the problem Newman exposed with the thesis is not that it treats theoretical statements as analytic or logically necessary, but that it represents them as *true* provided only that they are *satisfiable* in a domain that extends a model of the observable consequences.

The source of the difficulty Newman uncovered is the source of the structuralist thesis itself – namely the idea that only observation terms are fully understood – coupled with the contention that our partial under-standing of theoretical terms is sufficiently partial to justify the structuralist thesis. Notice that the difficulty is not with the *distinction* between obser-vation and theoretical terms but with the claim that the distinction has epistemological consequences for the nature of our understanding of the terms belonging to one or other vocabulary. The notion that the distinction is a distinction of epistemological importance is fostered by the dogma that unless it is accorded such significance, an important pillar of empiricism will collapse. But what the dialectic reveals is not a collapse of empiricism, but the failure of any view which is led by its understanding of empiricism to the structuralist thesis, and from that thesis to the equation of the mere satisfi-ability of a theoretical claim with its truth.

It is easy to overlook the dependence of Newman's problem on its philosophical context, and to fall into the view that it confronts us with a general and seemingly insoluble paradox. For in order to avoid the con-flation of the truth of a claim with its satisfiability, it is necessary to constrain the interpretation of the language of the claim so that its truth does not follow from the mere possibility of a correspondence of the sort employed in the proof of Newman's observation. How to do this for observation sentences appears transparent because the interpretation of their vocabulary is thought to rest ultimately on ostension. By contrast, it is characteristic of the theoretical vocabulary that it stands for things which transcend obser-vation. The theoretical vocabulary is therefore precluded from having its interpretation given by ostension. But if the interpretation of the theoretical vocabulary is not ostended, must it not be given by its satisfaction of a set of conditions – indeed, of the very conditions that are expressed by the "matrix" or quantified part of the Ramsey sentence of the theory? In this way one is led to conclude that although it is possible to establish the independence of the interpretation of the observation vocabulary from the truth of claims involving it, our epistemic situation prevents us from extending this independence to the interpretation of the theoretical vocabu-lary. It is important to recognize that this conclusion rests on a particular methodological diagnosis of how interpretations are established, one that necessarily assigns a more decisive role to experience in fixing the interpre-tation of the observation vocabulary than it assigns to its role in fixing the interpretation of the theoretical vocabulary.

An appropriate model for the interpretation of the theoretical vocabulary must start from the premise that the role of experience makes the inter-pretation of vocabulary equally provisional in both cases. It should acknowl-edge that while experience may guide us toward the interpretation of the theoretical and observation vocabularies, by itself it fixes the interpretation of neither. The schemes of interpretation to which we are led fluctuate over time and are susceptible to a process of revision at both the theoretical and observational levels. This process is never resolved by the prescription that a correct interpretation of the theoretical vocabulary is one that fulfills the matrix of the Ramsey sentence of the theory to which it belongs. The reason for this is normative and methodological: the distinction between the truth of a theoretical claim and its satisfiability is a condition of adequacy that we impose on any interpretive framework within which theory construction occurs. Empiricism demands that experience should make a rational contribution to our knowledge of the material world, but it allows that this requirement may be implemented in many ways. The view that the contribution of experience to the interpretation of the observation vocabulary is privileged by comparison with its contribution to the theoretical vocabulary represents only one development of empiricism, one which we have seen conflicts with a principle that is intuitively very compelling. I hope to have made the necessity of developing an alternative view which respects this intuition at least plausible. Anil Gupta's *Empiricism and Experience*[20] contains a framework that I believe can be adapted to fulfill this aim. By way of conclusion, let me indicate, if only very briefly, its general features and its bearing on our concerns.

Gupta's framework was developed for the purpose of presenting an account of the role of experience in empiricism which preserves the fundamental empiricist insight that experience not only makes a rational contribution to knowledge – what Gupta calls the "given in experience" – but is the supreme epistemic authority for our theoretical knowledge. Empiricists have traditionally held that it is *propositions* that comprise what is given by experience. There are deep skeptical difficulties to which this notion of the "propositional given" can lead when it is combined with a small number of very natural and compelling assumptions. These difficulties are explored at length in Gupta's book and will not be reviewed here. Rather, my concern is to indicate how the notion of the propositional given has functioned in the context of the theory of theories; how, on Gupta's account, it can be avoided without compromising the fundamental empiricist insight; and how Gupta's positive proposal might illuminate the issues which arise in connection with the structuralist thesis.

In the accounts of theories we have been discussing, the commitment to the propositional given is expressed by the epistemologically privileged role

---

[20] Gupta (2006).

these accounts accord observation sentences, and by the special epistemological status they accord the vocabulary with which observation sentences
are expressed. We have seen how for Russell the special epistemic status of
observation terms derives from our perceptual acquaintance with their
referents. But exactly *how* the epistemically privileged status of the observation vocabulary is spelled out is of less importance than the contrast all
three of the views we have described draw between observation and theoretical vocabulary. For each of these views, not only are propositions
expressed with theoretical vocabulary not part of the given in experience,
but because we are not given the meanings of their constituent theoretical
terms, theoretical statements are accorded an analysis that conforms to the
structuralist thesis.

The novelty of Gupta's approach lies in its retention of the fundamental
empiricist insight without the burden of the propositional given or of a class of
terms whose semantics is distinguished by special epistemic links to their
referents. The key idea underlying Gupta's positive proposal is the notion that
the given in experience is *hypothetical*: no proposition is given by experience
itself, but when experience is taken in conjunction with a "view" or "conceptual starting point" it yields hypothetical entitlements to judgments about
ourselves and our surroundings. These judgments ultimately figure in an
essential way in the articulation and defense of our commonsense and
theoretical conceptions of the world. As Gupta expresses it, the *logical form*
of the given is not *propositional* but *functional*. This means that what is given in
experience is not the entitlement to a proposition, but the entitlement to a
proposition *relative to a particular view*. In this respect, the given acts like a rule
of inference. For example, modus ponens is a functional relation between a
pair of propositions and the proposition which the application of the relation
yields as a value for that pair of propositions as argument. Rules of inference
give us *hypothetical entitlements*, which is precisely how, on Gupta's view, we
should conceptualize the given: relative to a view, the given yields a proposition to which we are hypothetically entitled but which we may later find
reason to revise. My suggestion is that we supplement this idea of hypothetical
entitlements to propositions with the proposal that relative to a view's conceptual apparatus, experience yields a hypothetical specification of the semantic content of the terms used to express the propositions to which we are
entitled: just as the judgments to which experience hypothetically entitles us
are subject to revision on the basis of experience, so also is the semantic
interpretation of the terms with which those judgments are expressed.

From the perspective of a reconstruction of theoretical knowledge that
distinguishes observation and theoretical terms, the revision process to

which all such hypothetical entitlements are subject involves both propositions which are the expression of observation sentences and propositions which are the expression of sentences that contain theoretical terms. Applying Gupta's framework to such a reconstruction, we find that experience informs the revision process without the assumption of a propositional given, and therefore without the assumption that the propositions given in experience are expressed in a vocabulary that is epistemologically distinguished. Within this framework the contribution of experience to the semantic properties of observation terms is methodologically on a par with its contribution to theoretical terms.

The emphasis this approach places on the notion of revision may suggest that it is just one among many proposals already canvassed during the protocol-sentence debate over the corrigibility of observation sentences and the possibility of formulating a successful empiricism without the assumption of a theory-neutral observation language.[21] But this debate did not address – nor did it seek to address – the paradoxes which call into question empiricism's ability to account for the things we take ourselves to know. The recognition of the corrigibility and theory-relative character of observation sentences certainly contributed to a liberalized and anti-foundationalist conception of justification. But by itself this recognition does nothing to solve the difficulties that stand in the way of sustaining the fundamental empiricist insight. Even Neurath's idea – that the revisability of what we take ourselves to know does not require the assumption of a fixed basis – leaves the status of the fundamental insight and the problem of articulating what is given in experience completely unresolved.[22]

Finally, it is important to note that the methodological neutrality between the observation and theoretical vocabulary which is essential to this elaboration of Gupta's framework is entirely compatible with the possibility that the psycho-linguistics of language comprehension may well pay special attention to the "observational" character of a term in an account of the process of its acquisition. And it is also compatible with the possibility that such a psychological theory may give very different accounts of our acquisition of "observation" and "theoretical" terms. What is at issue in the evaluation of different conceptions of theoretical knowledge is not the psychology of concept and belief acquisition, but the methodological bearing of experience on a theory's semantical and epistemological properties.

---

[21] See Uebel (2007) for an exhaustive study of this debate.
[22] There are of course importantly similar views which stress the concepts of revision and interdependence. On this, see Gupta (2006, pp. 104–5).

The conflation of the psychological and the methodological has long been recognized as one of the major defects of classical empiricism. What has been less clearly perceived is that this same conflation informs the significance empiricist accounts of theories attach to the observation–theory distinction. There is, to be sure, a complex and interesting story to be told about the psychology of how we come to understand terms which stand for things that are not accessible to our observation, and it may well be part of this story that a psychological theory of our understanding of such terms differs in important ways from its account of our understanding of terms which refer to things that are observable. But the presence of such psychological differences is independent of whether observation and theoretical terms make fundamentally different contributions to the truth of the sentences in which they occur. The importance of the revision theory of the methodology of concept acquisition and belief fixation sketched here is that it provides an alternative to the traditional empiricist elision of the distinction between methodological and psychological issues; at the same time the theory retains the decisive role empiricists have always accorded the contribution of experience to knowledge.

ACKNOWLEDGMENTS

I wish to thank Aaron Barth, James Ladyman, and Greg Lavers for their comments and questions on earlier drafts. Discussions with Hilary Putnam, Michael Friedman, and Anil Gupta on the topics dealt with here have been invaluable. I am indebted to the Social Sciences and Humanities Research Council of Canada and to the Killam Trusts for support of my research.

CHAPTER 8

Frege and the rigorization of analysis

8.1 INTRODUCTION

Paul Benacerraf, in his important paper, "Frege: the last logicist,"[1] raises a number of questions about the intellectual motivation for Frege's logicism. Benacerraf makes the correct point that when Frege's foundational interests are viewed in their mathematical context, they stand in sharp contrast with the logical empiricists' attempts to show the analyticity of arithmetic and, more generally, of all a priori knowledge. Any application of Frege's logicism to such an empiricist theory of the a priori is at best an unintended by-product of his foundational concerns. Benacerraf argues that assimilating Frege to a much later empiricism distorts his purpose to such an extent that if this were the point of logicism, we should not look to Frege for its intellectual origins. Frege's logicism comes at the end of another tradition, and that tradition is primarily a mathematical one. On Benacerraf's account, Frege's work is the culmination of the process of making rigorous the calculus and the theory of the reals. Frege sought to do for arithmetic what Cauchy, Bolzano, Weierstrass, Cantor, and Dedekind did for analysis: namely, to secure for it a "rigorous foundation." But what exactly was the goal of rigorization?

For Benacerraf, Frege's interest in rigor is driven by the problem of providing a proper justification for believing the truth of the propositions of arithmetic: "The propositions of arithmetic stand in need of proof" (p. 23) is Benacerraf's gloss on Frege's remark that "in mathematics a mere moral conviction, supported by a mass of successful applications, is not good enough" (*Grundlagen*,[2] sect. 1, quoted by Benacerraf, p. 23).

First published in *Journal of Philosophical Logic* 23 (1994): 225–46.
[1] Benacerraf (1981). All page references to Benacerraf are to this article.
[2] Throughout this chapter I use standard translations of Frege's writings. Quotations from *Grundlagen* are from Austin's translation. My citations of *Grundlagen* and *Begriffsschrift* – Frege (1884) and Frege (1879) – use the abbreviations *Gl* and *Bg*.

According to Benacerraf, both Frege and the nineteenth-century analysts viewed the process of rigorization as one of providing clear definitions and mathematical proofs where none had previously been given. This comports well with a number of Frege's statements, for example "it is in the nature of mathematics always to prefer proofs, where proofs are possible" (*Gl*, sect. 2, quoted by Benacerraf, p. 23), and "the rigor of the proof remains an illusion . . . so long as the definitions are justified only as an afterthought, by our failing to come across any contradiction" (*Gl*, p. xxiv, quoted by Benacerraf, pp. 22–3).

On Benacerraf's account of the goal of rigorization, Frege's foundational investigations are epistemological in the standard sense that the questions and meta-questions they address surround the *justification* of mathematical propositions for the purpose of ensuring that we will not "in the end encounter a contradiction which brings the whole edifice down in ruins" (*Gl*, p. xxiv, quoted by Benacerraf, p. 23). It should therefore be clear that the question Benacerraf is raising is not exhausted by the question whether Frege's foundational interests were mathematical or philosophical: they were obviously *both*. Benacerraf is also claiming that our understanding of Frege's foundational interests is invariably distorted when these interests are viewed from the perspective of a later philosophical school; although I do not address it directly, I believe that what I have to say is supportive of this broad interpretive claim.

It would be foolish to deny that the goals of cogency and consistency were an important part of the nineteenth-century interest in rigor. And it is easy to find quotations from Frege which show that he sometimes at least wrote as if he took his task to include securing arithmetic – in the broad sense of 'arithmetic' which includes both the arithmetic of natural numbers and the theory of the reals – against contradiction. But it can be questioned just how far skeptical worries about the consistency or cogency of mathematics, generated perhaps by a certain incompleteness of its arguments, were motivating factors for Frege's logicism or for the other foundational investigations of the period. This is a large historical issue which I will not attempt to resolve here. Rather, I will restrict myself to identifying another, largely neglected, component of Frege's concern with rigor. This component not only has some intrinsic interest, it also elucidates his views on the nature and significance of intuition in mathematical proof as well as his conception of his mathematical and foundational accomplishments, especially those of *Begriffsschrift*.

Frege's concern with rigor is closely tied to his rejection of intuition in arithmetical reasoning. This interconnection has not, I think, been fully understood; at any rate, its importance for our appreciation of the scope and justification of his logicism has not been sufficiently appreciated. Since, on the view to be presented here, Frege inherited his concern with rigor from the tradition in analysis initiated by Cauchy and Bolzano and carried forward by Weierstrass, Cantor, and Dedekind, what I have to say concerning Frege is intended to apply to their foundational interests as well. Indeed taken solely as a historical thesis about the rigorization of analysis, my remarks are neither novel nor contentious, having been well emphasized by the historians.[3] But to my knowledge, with the exception of Michael Dummett,[4] no one has applied these ideas to Frege in the way I propose here.[5]

8.2 RIGOR AND SPATIO-TEMPORAL INTUITION

Shortly after the publication of *Begriffsschrift* Frege wrote a long study of its relationship to Boole's logical calculus.[6] The paper carries out a detailed proof (in the notation of *Begriffsschrift*) of the theorem that the sum of two multiples of a number is a multiple of that number. In addition to the laws and the (explicit and implicit) rules of inference of *Begriffsschrift*, Frege appeals only to the associativity of addition and to the fact that zero is a right identity with respect to this operation. He avoids the use of mathematical induction by applying his definition of *following in a sequence* to the case of the number series; 'following in a sequence' is Frege's expression for the ancestral of the relation which orders the sequence.[7] The paper also includes definitions of a number of elementary concepts of analysis (again in the notation of *Begriffsschrift*). In the course of this discussion Frege presents an illuminating account of the considerations that shaped his logic. Speaking

---

[3] Especially Youschkevitch (1976), Grabiner (1981), and Bottazzini (1986).
[4] See especially Dummett (1991a). *Added in 2012*: The paper this chapter reprints was written before Dummett's book appeared.
[5] Coffa (1982) stresses Frege's affinity with the tradition in mathematical analysis, especially his affinity with Bolzano. Coffa's perspective is, however, somewhat different from mine since on his account, pure intuition was "a cancer that had to be extirpated in order to make mathematical progress possible" (Coffa 1982, p. 686). I hope to show that whatever its role in the development of the calculus, this was not a factor in Frege's elimination of intuition from the proofs of various arithmetical principles concerning the natural numbers.
[6] Frege (1880/1). When citing Frege's works I have inserted the page numbers of the German original (or standard German edition) in brackets, immediately after those of the standard English edition in which the work appears in translation. Otherwise quotations are identified by the section numbers of the works from which they are drawn.
[7] Sometimes the translations I quote use 'following in a series.' Frege's definition of the ancestral is reviewed in Section 8.3 below.

of the choice of proposition given a "step by step" or "gap-free" proof in *Begriffsschrift*, Frege writes:

So as not perhaps to overlook precisely those transformations which are of value in scientific use, I chose the step by step derivation of a sentence which, it seems to me, is indispensable to arithmetic, although it is one that commands little attention, being regarded as self-evident. The sentence in question is the following:

> If a series is formed by first applying a many–one operation to an object (which need not belong to arithmetic), and then applying it successively to its own results, and if in this series two objects follow one and the same object, then the first follows the second in the series, or vice versa, or the objects are identical.

This is the proposition (133) whose proof concludes *Begriffsschrift*. I will refer (somewhat loosely) to what this proposition establishes as the connectedness of the ancestral. I say "somewhat loosely" because the condition imposed on "connected elements," namely, that they follow one and the same element, is not part of the standard definition of connectedness, and because connectedness, in this extended sense, holds only of the ancestral of *many–one* relations.

Frege continues,

I proved this sentence from the definitions of the concepts of following in a series, and of many-oneness by means of my primitive laws. In the process I derived the sentence [(98)] that if in a series one member follows a second, and the second follows a third, then the first follows the third. Apart from a few formulae introduced to [accommodate] Aristotelian modes of inference, I only assumed such as appeared necessary for the proof in question.

These were the principles which guided me in setting up my axioms [*ursprünglischen Sätze*] and in the choice and derivation of other sentences. It was a matter of complete indifference to me whether the formula seemed interesting or to say nothing. That my sentences have enough content, in so far as you can talk of the content of sentences of pure logic at all, follows from the fact that they were adequate for the task.[8]

It seems not to have been noticed – at any rate it has not been sufficiently emphasized – that neither in this paper nor in *Begriffsschrift* does Frege suggest that connectedness is not correctly regarded as self-evident, or that without the proof of *Begriffsschrift* one might reasonably doubt the connectedness of the ancestral, and with it the justification of all those

---

[8] All quotations in this paragraph are from Frege (1880/1, p. 38 [42–3]). I do not know whether it is generally recognized that the first complete axiomatization of first-order logic was determined by reflection on what the proof of the connectedness of the ancestral would require, but it is a remarkable fact that the discovery of a gap-free proof of exactly this proposition should have sufficed to generate the first-order fragment of Frege's logic.

propositions which require it. The point is rather that without a gap-free proof of this proposition, or of its specialization to the number series, one might be misled into thinking that arithmetical proofs which depend on connectedness import considerations based on "intuition." As Frege puts the matter in the introductory paragraph to Part III of *Begriffsschrift*:

Throughout the present example [i.e. the proof of 133, which occupies the whole of Part III] we see how pure thought, irrespective of any content given by the senses or even by an intuition *a priori*, can, solely from the content that results from its own constitution, bring forth judgements that at first sight appear to be possible only on the basis of some intuition.[9]

This point is also made in *Grundlagen* when, near the end of the work (*Gl*, sects. 90–1), Frege comments on *Begriffsschrift*. Frege is quite clear that the difficulty with gaps or "jumps" in the usual proofs of arithmetical propositions is not that they may hide an unwarranted or possibly false inference, but that their presence obscures the true character of the reasoning:

In proofs as we know them, progress is by jumps, which is why the variety of types of inference in mathematics appears to be so excessively rich; the correctness of such a transition is immediately self-evident to us; whereupon, since it does not obviously conform to any of the recognized types of logical inference, we are prepared to accept its self-evidence forthwith as intuitive, and the conclusion itself as a synthetic truth – and this even when obviously it holds good of much more than merely what can be intuited.

   On these lines what is synthetic and based on intuition cannot be sharply separated from what is analytic . . .

   To minimize these drawbacks, I invented my concept writing. It is designed to produce expressions which are shorter and easier to take in . . . so that no step is permitted which does not conform to the rules which are laid down once and for all. It is impossible, therefore, for any premiss to creep into a proof without being noticed. In this way I have, without borrowing any axiom from intuition, given a proof of a proposition [namely, 133] which might at first sight be taken for synthetic.

There is an understandable tendency to pass over such passages, because of the difficulty of the Kantian concept of an a priori intuition. But I think it is possible to understand Frege's thought without entering into a detailed investigation of this concept. It suffices to recall that for the Kantian mathematical tradition of the period, our a priori intuitions are of space and time, and the study of space and time falls within the provinces of geometry, kinematics, and perhaps mechanics. It then follows that the dependence of a basic principle of arithmetic on some a priori intuition

---

[9] *Bg*, p. 55 (sect. 23).

would imply that arithmetic lacks the autonomy and generality we associate with it: to establish its basic principles, we would have to appeal to our knowledge of space and time, and then arithmetical principles, like the connectedness of the ancestral and mathematical induction, would ultimately come to depend on our knowledge of spatial and temporal notions for their full justification. Frege's point in the passages quoted is that even if it were possible to justify arithmetical principles in this way, it would be a mistake to suppose that such an external justification is either necessary or appropriate when a justification that is internal to arithmetic is available. The search for proofs, characteristic of a mathematical investigation into foundations of the sort Frege was engaged in, is not motivated by uncertainties concerning basic principles or their justification, but by the absence of an argument which establishes their autonomy from geometrical and kinematical ideas. And autonomy is important since the question of the *generality* of the principles is closely linked to the issue of their autonomy and independence from our knowledge of space and time.[10]

On this "minimalist" reading of the allusion to Kantian intuition, the rigor of *Begriffsschrift* is required in order to show that arithmetic has no need of spatial or temporal notions. Indeed, Frege's concern with autonomy and independence suffices to explain the whole of his interest in combating the incursion of Kantian intuition into arithmetic. In this respect, Frege's intellectual motivations echo those of the nineteenth-century analysts who sought to free the calculus and the theory of the reals from any dependence on the sciences of geometry and kinematics. Thus as early as 1817 Bolzano wrote: "the concepts of *time* and *motion* are just as foreign to general

[10] The view presented in the text may perhaps be further clarified if it is contrasted with Kitcher (1979, p. 245), who writes:

[The reasoning of mathematicians who rely on spatio-temporal intuition] leaves much to be desired. Transitions from one proposition to another are made in huge jumps so that, even if the premises were known for certain, it might be reasonable to doubt the conclusion. Existing proofs codify the slipshod practices of mathematicians. They should be replaced by genuine proofs . . . which proceed by steps invulnerable to doubt.

Kitcher, like Benacerraf, holds that a jump is a source of doubt. For both Kitcher and Benacerraf the purpose of rigor – of gap-free proofs – is the removal of such sources of possible doubt. Kitcher therefore agrees with Benacerraf's view of Frege's foundational interests. They differ over the broader interpretive claim, with Kitcher holding that the epistemological concerns they both attribute to Frege fall within a traditional, quasi-psychologistic, theory of knowledge and mathematical proof. But this claim is highly questionable if, as the discussion in the text shows, Kitcher's (and Benacerraf's) particular account of the goal of rigor is unsupported. In my view, Michael Dummett is the only commentator to have gotten Frege's view of the significance of gaps exactly right. See especially Dummett (1982, pp. 24–5).

mathematics as the concept of *space*."[11] And over fifty years later Dedekind was equally emphatic:

For our immediate purpose, however, another property of the system 𝕽 [of real numbers] is still more important; it may be expressed by saying that the system 𝕽 forms a well-arranged domain of one dimension extending to infinity on two opposite sides. What is meant by this is sufficiently indicated by my use of expressions borrowed from geometric ideas; but just for this reason it will be necessary to bring out clearly the corresponding purely arithmetic properties in order to avoid even the appearance [that] arithmetic [is] in need of ideas foreign to it.[12]

It must be conceded that the issues raised by the foundations of analysis are rather more varied than they are in the case of the arithmetic of natural numbers; in the case of the real numbers, the usual explanation of the purpose of rigor, as a guarantee of cogency and a hedge against inconsistency and incoherence, while not a complete account of the matter, is certainly part of the story. But when we turn to Frege's primary concern about the domain of intuition, namely, the arithmetic of natural numbers, skepticism simply plays no role in any of his arguments against its use. We just do not find Frege rejecting intuition for fear that it is a potentially faulty guide to truth. Those few passages which suggest otherwise are invariably concerned with arithmetic in the broad sense which includes real analysis. When we factor arithmetic by real analysis we also factor out doubts about cogency and consistency. What is left is a concern with autonomy and independence of the sort expressed by Bolzano and Dedekind in the passages quoted earlier. Thus while Benacerraf is certainly right to emphasize the mathematical context of Frege's foundational investigations, he has, I think, missed a distinctive feature of those investigations.[13]

We may briefly summarize our discussion of the historical situation as follows. The interest in rigor has both philosophical and mathematical aspects. To begin with, we require proofs internal to arithmetic. This idea is always presented as a prohibition – a prohibition against the incursion of spatial and temporal notions into demonstrations of the propositions of arithmetic. Finding proofs which are free of such notions is the mathematical aspect of rigorization. The motivation underlying the foundational

---

[11] Russ (1980, p. 161).   [12] Dedekind (1872, p. 5).

[13] In the case of real analysis, Frege emphasized the importance of establishing its independence from spatio-temporal intuition as early as his 1874 qualifying essay: "there is . . . a noteworthy difference between geometry and arithmetic in the way in which their fundamental principles are grounded. The elements of all geometrical constructions are intuitions and geometry refers to intuitions as the source of its axioms. Since the object of arithmetic does not have an intuitive character, its fundamental propositions cannot stem from intuitions either" (Frege 1874, pp. 56–7 [1]).

interest in rigor has a philosophical component. It is not, as Benacerraf has correctly observed, the program of showing that the truths of arithmetic are true in virtue of meaning, or conventions of language. Benacerraf is also right to insist that it is not usefully viewed from the perspective of any of the familiar subsequent philosophical positions. But the foundational interest in rigor is not a concern with justification in the face of doubt – mathematical or skeptical. Rather, Frege's epistemological interests are broadly architectonic: the philosophical dimension to his foundational program is to establish the independence of our knowledge of arithmetical principles from our knowledge of spatial and temporal notions. This is so basic a feature of the tradition in foundations to which Frege belongs that Dedekind, for example, actually explains his logicism in terms of it:

> In speaking of arithmetic (algebra, analysis) as part of logic I mean to imply that I consider the number concept entirely independent of the notions or intuition of space and time, that I consider it an immediate result from the laws of thought.[14]

8.3 SPATIO-TEMPORAL INTUITION AND ABSTRACTION

While it is clear how, for example, the problem of characterizing continuity might suggest the necessity of spatio-temporal ideas in the theory of the reals, it is far from obvious how, in the case of the arithmetic of natural numbers, an arithmetical principle might be thought to require spatio-temporal notions. A suggestion of how such a view might arise and come to seem inevitable is contained in Frege's discussion of attempts by Thomae, and others, to connect the possibility of counting with position in a spatio-temporal series. The grouping to which this section belongs is entitled "Attempts to overcome the difficulty." The difficulty referred to is formulated in the previous section (sect. 39) as an objection to the Euclidean idea that a natural number is an agglomeration, either of units or of objects, arrived at by a process of abstraction. The problem seems to be this: if we abstract from the individuality of the elements of a collection, we will not have several objects, but will have instead only one. So in this case no collection can be identified with a number greater than one. But if we do not abstract over their individual differences, each collection of (say) four objects will be a different number four.[15] For someone committed to the

---

[14] Dedekind (1888, p. 31); this is taken from the first page of the first-edition preface to "The nature and meaning of numbers." (Notice Dedekind's parenthetical clarification of the scope of arithmetic.) For a discussion of Dedekind and Cantor along related lines, see Hallett (1990, pp. 228–33).
[15] Cf. *Gl*, sect. 39: "If we try to produce the number by putting together different distinct objects, the result is an agglomeration in which the objects contained remain still in possession of precisely those

idea that the numbers are arrived at by abstraction, it is necessary to seek some constraint on how far the process of abstraction can be carried. This is evidently the idea behind Thomae's suggestion that numbers are associated with the positions of a spatio-temporal series:

If we consider a set of individuals or units in space and number them one after the other, for which time is necessary, then, abstract as we will, there remains always as discriminating marks of the units their different positions in space and in the order of succession in time. (quoted by Frege: *Gl*, sect. 40)

Thus one source of the persistence of the idea that our knowledge of the arithmetic of natural numbers depends on our concepts of space and time is an internal difficulty in the abstractionist theory of the formation of simple numerical concepts like counting. The theory is committed to the process of "subtracting" properties by its account of the difference between numbers and ordinary collections: a collection "becomes a number" when, by the "removal" of properties from its members, it becomes a collection of units. But without some restriction on how far such subtractions of properties may be carried, the abstractionist theory of number is threatened with the absurdity that there is only one number. Spatio-temporal position provides the necessary constraint.

Notice that when abstraction is used to explain the formation of the ordinal notion of number, what for us is a metaphorical use of 'position,' as, for example, when we speak of position in a series, is not at all metaphorical for an abstractionist like Thomae. Lacking a *general* notion of order, Thomae's understanding of ordinal position is *literally* tied to the example of a spatio-temporal series. This observation did not escape Frege's attention, for when in *Grundlagen* he recalls the *Begriffsschrift* definition of the ancestral, he is careful to remark on the fact that his definition is not restricted to a spatial or a temporal series:

It will not be time wasted to make a few comments on this. First, since the relation [whose ancestral has been defined] has been left indefinite, the series is not necessarily to be conceived in the form of a spatial and temporal arrangement, although these are not excluded.

Frege's account of the ancestral proceeds schematically: he takes a fixed, but arbitrary, binary relation *f*, and shows how to form its ancestral. The

properties which serve to distinguish them from one another; and that is not the number. But if we try to do it in the other way, by putting together identicals, the result runs perpetually together into one and we never reach a plurality." For a discussion of Cantor's abstractionism, see Hallett (1984, Chapter 3). For a discussion of Husserl's abstractionism, see Dummett (1991c, Chapter 2), "Frege and the paradox of analysis."

account is general ("the relation has been left indefinite"), since, aside from the assumption that f is a binary relation, no further conditions are imposed on f. Frege's definition is easily presented in three simple steps.[16]

Suppressing explicit reference to the relation f, we abbreviate 'property F is hereditary in the f-sequence' by 'Her(F)', and define it by the condition

$$\forall d \forall a (\, dfa \,\&\, Fd \text{ only if } Fa),$$

given by Proposition 69 of *Begriffsschrift*. For 'after x, property F is inherited in the f-sequence' we write 'In(x, F),' and define this as

$$\forall a (xfa \text{ only if } Fa).$$

For 'y follows x in the f-sequence' – the ancestral of f – we write 'xf^*y.' With the above abbreviations, Frege's definition of the ancestral becomes:

$$xf^*y =_{Df} \forall F (\text{Her}(F) \,\&\, \text{In}(x, F) \text{ only if } Fy).$$

What is striking about Frege's definition comes out most clearly when it is compared with characterizations to which it turns out to be equivalent. The intuitive or pre-analytic characterizations are very heavily imbued with number-theoretic and spatio-temporal terminology, as is apparent when, for example, we say that y follows x in the f-sequence if, beginning with x, y is reached after only finitely many applications of the procedure f. Frege's definition avoids this, and his remark in *Grundlagen* to this effect is clearly justified.

8.4 FREGE'S THEORY OF SEQUENCES

Part III of *Begriffsschrift* is entitled "Some topics from a general theory of sequences." For a modern reader, the work deals with topics from the theory of binary relations. The modern analogue of a sequence in Frege's sense is simply a nonempty set X, together with a binary relation f on X, i.e. $f \subseteq X \times X$. However we will see that something conceptually important is missed if we identify f with a set of ordered pairs.

The major conceptual innovation of Part III is the definition of the ancestral, and the principal result of Part III is the proof of the connectedness of the ancestral. Frege's proof makes extensive use of an implicit substitution rule which allows him to substitute formulas for property

---

[16] See Boolos (1985). All references to Boolos are to this paper; page numbers refer to its original place of publication.

variables. It is well known that the substitution rule is equivalent to a comprehension scheme,

$$\exists P \forall x (Px \text{ if and only if } A(x))$$

(where P is not free in A), and it therefore constitutes an assumption concerning property existence.[17] The properties required by Frege's proofs are expressed by formulas of the form 'mf^*x' ('m' a constant) and thus depend on the definition of the ancestral. Since the definition of the ancestral of f contains a second-order quantifier, properties defined by such formulas are defined impredicatively. Frege understands the formula, 'mf^*x' to express the property of following m in the f-sequence. (See, for example, *Bg*, p. 80, sect. 31.)

It has long been known that properties of this form are closely related to Dedekind's notion of a *chain*.[18] Recall that for Dedekind a *chain* is a pair $\langle X, f \rangle$ consisting of a nonempty set X and a one–one function f, such that

$$X \subseteq dom(f)$$

and

$$f[X] \subseteq X.$$

For an element $a \in X$, Dedekind writes 'a_0' for the *chain* of a, and puts

$$a_0 =_{Df} \cap \{X : \langle X, f \rangle \text{ is a } chain \ \& \ a \in X\}.$$

To state the exact connection between Dedekind's chains and Frege's properties, define the *weak ancestral* $f_=^*$ of f by

$$x f_=^* y =_{Df} x f^* y \ or \ x = y.$$

Then, on the assumption that f is one–one, a_0 is just the extension of the property $af_=^* y$, defined by the weak ancestral of f. Because of their connection with chains, such properties evidently possess a rich mathematical content.

It is clear that Frege's focus in *Begriffsschrift* is on principles that the number sequence (i.e. the set of natural numbers under the successor

---

[17] See Boolos (1985, p. 337) for a proof.

[18] Presented in sects. 37 and 44 of Dedekind (1888), "The nature and meaning of numbers." As remarked in Dedekind (1890), the notion of a chain does for Dedekind's analysis of the number sequence what the ancestral does for Frege's general theory of sequences. See Wang (1957) for a discussion of Dedekind (1890).

relation) shares with other sequences. This suggests a natural division in his logicism between this early concern with principles, such as the connectedness of the ancestral and the reconstruction of inductive reasoning,[19] and his later concern with numbers as "objects," the concern in *Grundlagen* and *Grundgesetze* with the definition of the natural numbers and the proof of their infinity. How should we assess the success of Frege's account of these principles?

There is an influential line of criticism, originating with Poincaré,[20] but most forcefully presented by Charles Parsons.[21] Parsons is concerned with the Fregean analysis of the predicate, 'is a natural number,' but if his objection is successful, exactly the same difficulties confront Frege's definition of the ancestral. The objection focuses on the use of induction in the proof of the adequacy of the definition of the ancestral, where an adequacy proof is understood as one which establishes that the definition succeeds in defining a concept that is equivalent to the intuitive or pre-analytic one. Let 'xf^*y' denote the reconstructed ancestral, and let 'y follows x in the f-sequence' denote the intuitive ancestral. Then to establish adequacy, we proceed inductively by showing that xf^*y if, and only if, y follows x in the f-sequence when y is 1 "f-step" from x, and if the equivalence holds when y is n f-steps from x, it also holds when y is $n + 1$ f-steps from x. We conclude that the equivalence holds however finitely many f-steps separate x and y. But proceeding in this way, we rely on our pre-analytic understanding of the ancestral and on the *underived* principle of mathematical induction. The claim that the definition of the ancestral yields a *foundation* for inductive reasoning cannot be sustained, Parsons and Poincaré argue, if, in order to establish the adequacy of the definition, we must fall back on intuitive reasoning.

In the postscript to the reprinting of his original paper, Parsons suggests that this argument is inconclusive, since it is open to the Fregean to use the derived principle of induction in the proof of adequacy. Parsons argues that the circularity this seems to introduce is similar to the "circularity" that accompanies standard metalogical justifications of formal reconstructions, in which case such a use of the derived principle of induction may be acceptable after all. But there is a simpler and more decisive objection to the Parsons–Poincaré line of criticism. If the cogency of intuitive reasoning is

---

[19] In *Bg* (sects. 24 and 26) Frege proves a generalized form of mathematical induction, which he formulates as follows: if x has the property F that is hereditary in the f-sequence, and if y follows x in the f-sequence, then y has the property F (*Bg*, p. 62, Proposition 81).

[20] Who urged it against Couturat; there is no indication that Poincaré ever read Frege. For a discussion and references, see Goldfarb (1989).

[21] Parsons (1964).

not at issue, then it is a wholly pragmatic question whether we should justify the adequacy of the definition of the ancestral by appealing to the intuitive or the derived principle of induction. Evidently, to convince Poincaré, Frege would do well to use underived induction. But this tells us rather more about Poincaré than about Frege's account of the principle of induction. Poincaré-type objections have succeeded in gaining a following only because the foundational program they are directed against has been misrepresented as a naive form of foundationalism.

More recently, George Boolos has raised a question about the mathematical development of Part III, and in particular about Frege's use of substitution in the proof of 133. Boolos puts the point by having his Kantian interlocutor ask whether it is not "an intuition of precisely the kind Frege thinks he has shown unnecessary that licenses [the inference to properties like x follows m in the f-sequence]" (Boolos [1985], pp. 337–8). To crystallize the point, we should distinguish the general theory – the case which leaves f indefinite – from its application to the natural number sequence. While both cases suggest questions about the bearing of intuition on assertions of property existence, the questions are rather different. Once the coherence of investigating an arbitrary but fixed f is admitted, it is difficult to see how any question can be raised about the intuitive – i.e. spatio-temporal – character of the properties of the elements it relates, or about the spatio-temporal character of the knowledge on which the inference to such properties rests. This of course leaves open the possibility that in some other sense of 'intuition,' intuition may have a role to play in the general case. Nevertheless, a Kantian can make the point only by denying the coherence of our understanding of a standard practice. I think it is clear that Frege can safely ignore such a Kantian, and assume that the notion of a property of elements of an arbitrary but fixed sequence is not necessarily tied to spatio-temporal intuition.

What of the Kantian who allows the general theory outlined in Part III, but questions whether its application to the sequence of natural numbers is capable of proceeding without spatio-temporal intuition? This possibility raises a reasonable concern, and it is evident that addressing it will require considerations that go beyond what is explicitly presented in *Begriffsschrift*.

In Frege's later work, most explicitly in *Function and Concept* which appeared twelve years after *Begriffsschrift*, properties are assimilated to Fregean concepts, and concepts are then treated as a special type of function. But the assimilation of properties to functions is just an elaboration of *Begriffsschrift*'s rejection of the subject–predicate analysis of sentences in favor of function-argument (or function-variable) notation. And it is clear that both developments were the result of reflection on the evolving understanding of functions in mathematical analysis.[22] Frege in effect recognized that a traditional notion of logic, namely that of a *property of numbers*, could be subsumed under a slight generalization of the mathematical concept of an arithmetical function. He further recognized that the arithmetical properties required for the development of arithmetic are contained in the class of characteristic functions, and that no *other* notion of property is needed for the development of arithmetic.

An important consequence of the nineteenth-century rigorization of analysis was the vindication of the idea that a real-valued function is simply a many–one correspondence between real numbers. As early as the 1830s Dirichlet and Lobachevsky presented clear and unambiguous statements of the modern notion of a *continuous* function from this point of view. But the idea that the general notion of an arithmetical function could only be captured by the concept of an arbitrary correspondence – and that it might be mathematically useful so to capture it – is probably best credited to Riemann's work of 1854.[23] A decisive step in this development was taken when it was recognized that the concept of a real-valued function is not constrained by the notion of a possible trajectory of a classical particle, or that of a piecewise continuous curve – steps that clearly removed the

---

[22] *Function and Concept* (1891) is of course replete with allusions to analysis. And the subtitle of *Begriffsschrift – A Formula Language, Modeled upon that of Arithmetic, for Pure Thought* – is an allusion to Frege's appropriation of the function-variable notation of analysis, something that is easily missed if the use of 'arithmetic' in the subtitle is understood too narrowly. The influence of analysis is explicitly acknowledged in sect. 10, where Frege writes: "the concept of function in analysis, *which in general I used as a guide*, is far more restricted than the one developed here" (*Bg*, p. 24, emphasis added). Frege's point is that *Begriffsschrift*'s characterization of a functional expression is general enough to subsume *both* the usual notion of a functional expression for a function of first level *and* that of a functional expression for a function of second level, so that *Begriffsschrift*'s notion of a functional expression – and therefore its concept of a function – is less "restricted" than what one finds in analysis. The fact that Frege's celebrated explanation of a function is actually a characterization of a functional expression is sometimes interpreted as showing that *Begriffsschrift* confuses functions with functional expressions. For two examples, see Bell (1990) and Simons (1988). But this is just a mistake. The functions of the later work are termed 'single-valued procedures' (*eindeutige Verfahren*) in *Begriffsschrift*. And Frege certainly recognizes the distinction between single-valued procedures and their linguistic expression.

[23] Thus Bottazzini remarks that "if for Dirichlet ... the generality of the definition [of a real-valued function] did not go along with a consistent practice in the study of equally 'general' functions, the opposite is true for Riemann. In his thesis (*Habilitationsschrift*) of 1854 ... he revealed to the mathematical world a universe extraordinarily rich in 'pathological' functions" (Bottazzini 1986, p. 217). Bottazzini goes on to remark that when one looks at the historical development, "the modern concept of a function of a real variable, in its full generality, began to emerge in mathematical practice only towards the end of the 1860s as Riemann's writings became more widely known" (*ibid.*). The later questioning of the notion of an arbitrary correspondence by the French school of quasi-intuitionists, toward the end of the nineteenth century, shows that the story is an intricate one. An illuminating discussion of the importance of the development of the function concept for philosophy of science is presented in Wilson (1993).

concept from its earlier dependence on geometry and kinematics. Since dependence on geometry and kinematics pretty much exhausts what for Frege is meant by dependence on a priori intuition, it is reasonable to assume on the basis of this development that the notion of an arithmetical property, whether of real or of natural numbers, owes nothing to Kantian intuition. It follows that the techniques developed in *Begriffsschrift* yield successful derivations of mathematical induction and the structural properties of the ancestral of the successor relation, and that all this is achieved without appealing to intuition, in the then generally accepted understanding of what such an appeal would mean.

The strategy behind this defense of Frege's account of various principles, is slightly unusual: we have exploited a feature of the *mathematical* development of the notion of an arithmetical property in order to defend a *logicist* thesis regarding the arithmetic of natural numbers. The thesis defended is the relatively weaker claim – weaker than the thesis that arithmetic is reducible to logic – that certain fundamental arithmetical principles do not depend on spatio-temporal intuition, something that the mathematical development sustains. But there is considerably more to the mathematical development of the function concept and of the notion of an arithmetical property than their removal from the domain of spatio-temporal intuition. Our defense is successful to the extent that it does not depend on the characterization of a function as a many–one correspondence, or that of an arithmetical property as a *set* of numbers. In order to subsume the positive characterization of these notions, Frege must appeal to the value ranges of functions and the extensions of concepts. To what extent does *Begriffsschrift*'s account of the principles avoid Frege's theory of extensions and value ranges together with the difficulties associated with their introduction?

It is well known that, taken by itself, Frege's theory of concepts is consistent, concepts being stratified; it is only the combination of concepts plus extensions that leads to paradox. Moreover, the stratification of concepts is naturally extended to the properties of *Begriffsschrift* and yields a consistent interpretation of that work, since there is nothing in the formal apparatus of *Begriffsschrift* that requires the association of properties with extensions. This is an important point of difference between *Begriffsschrift* and *Grundgesetze*;[24] extensions are nowhere mentioned in the earlier work.

---

[24] Frege (1893/1903).

In any case, they appear to play no fundamental role in its exposition.[25] It is only when Frege sets out to prove the infinity of the natural number sequence that a role for extensions becomes a prominent feature of his account. However, these considerations do not show that the connection between properties and their extensions may be ignored. Without an account of this connection, our grasp of the properties and relations of *Begriffsschrift* is incomplete. For example, what are we to say of the connection between the properties $af_={}^*y$ and Dedekind's chains? To express the connection, we, of course, would appeal to an axiomatic development of the set concept. But for Frege this is highly problematic, since it would bring his whole program to rest on yet another "special science." If Frege did not consider this consequence, this was because he assumed Basic Law v of *Grundgesetze* would provide the required representation of properties and relations by their extensions. What strikes one as the insurmountable obstacle to Frege's logicism, even to the complete implementation of the modest program Frege set himself in *Begriffsschrift*, is its failure to derive extensions from properties: the sorry history of Basic Law v.

ACKNOWLEDGMENTS

I wish to thank Robert Butts, John Corcoran, Michael Dummett, Michael Friedman, Michael Hallett, and Mark Wilson, for their considered comments on earlier drafts, and their continued encouragement. I am indebted to the Social Sciences and Humanities Research Council of Canada for financial support.

---

[25] An impression that is borne out by a remark in the correspondence where, in a letter to Jourdain (dated September 23, 1902) which discusses *Begriffsschrift*, Frege writes: "This distinction between a heap (aggregate, system) and a class, which had not perhaps been drawn sharply before me, I owe, I believe, to my conceptual notation, although you will not perhaps discover any trace of it in reading my little work. Some of the reasonings that occurred in writing it may well have left no trace in print." The letter is collected in McGuinness (1980, p. 73).

CHAPTER 9

The philosophical basis of our knowledge of number

After briefly sketching what I take to have been the major intellectual motivation for Frege's logicism, I will consider three topics arising from its presentation in *Grundlagen*[1]: (i) Frege's contextual definition of the cardinality operator and his derivation from it of the infinity of the number sequence – what has come to be known as *Frege's theorem*; (ii) Frege's deployment of the context principle as the basic tool in his analysis of the central question of *Grundlagen*, namely, the question: 'How are numbers given to us if we have neither experience nor intuitions of them?'; and (iii) the degree of success that we may, with hindsight, see Frege to have achieved, both with respect to this question and with respect to other aims of his logicist program. The chapter concludes with an appendix on the neo-Fregean interpretation of the contextual definition.

9.1

A view of Frege's philosophy of arithmetic that has gained wide currency is that it marks the culmination of the process of making rigorous the calculus and the theory of the real numbers; briefly put, Frege sought to do for arithmetic what Cauchy, Bolzano, and others did for analysis: namely, to secure for it a "rigorous foundation." But until the interest in rigor is spelled out, this really tells us very little.

On the view I favor, Frege's concern with rigor, like that of Dedekind and other mathematicians of the period, was closely tied to his rejection of intuition in arithmetical reasoning. Numerous passages from *Begriffsschrift* and *Grundlagen* can be cited to show that Frege's difficulty with gaps or

---

First published in *Noûs* 32 (1998): 481–503.
[1] Frege (1884). My citations of Frege's *Grundlagen* and his *Begriffsschrift* – Frege (1879) – use the abbreviations *Gl* and *Bg*, respectively.

jumps in the usual proofs of arithmetical propositions was not that they might hide an unwarranted or possibly false inference, but that their presence obscured the true character of the reasoning: a gap-free demonstration from general truths is one which does not depend on intuition; and the existence of just one such gap-free demonstration suffices to establish this.[2] Freedom from intuition was important, not because of some doubt about the cogency of "intuitive" reasoning. The difficulty for Frege was that if our proofs are intuitive in the Kantian sense of requiring an appeal to our knowledge of space and time, then arithmetical principles, like the structural properties of the ancestral of the successor relation, or the principle of mathematical induction, would ultimately come to depend upon our knowledge of spatial or temporal notions for their full justification. Frege's point was that even if it were possible to justify arithmetical principles in this way, it would be a mistake to suppose that such an external justification is either necessary or appropriate if a justification that is internal to arithmetic is available. In short, the search for gap-free proofs, characteristic of an investigation into foundations of the sort Frege was engaged in, was not motivated by uncertainty or skepticism concerning basic principles or their justification, but by the absence of an argument establishing their *autonomy* from geometrical and kinematical ideas. And autonomy is important because the question of the generality of the principles of arithmetic is so closely linked to the issue of their independence from our knowledge of space and time. Although the justification for rigor was almost always presented in terms of a prohibition of spatial and temporal notions from the demonstrations of the propositions of arithmetic, it had an important positive justification: namely, the desire to recover the evident generality of arithmetical reasoning – a feature to which Frege frequently returns in *Grundlagen*. I will have more to say about this perspective on Frege's aims toward the end of the chapter.

9.2

Let me begin my discussion by considering Frege's implementation of that part of his program which centers on the contextual "definition" of the cardinality operator. This definition, presented in *Gl* (sect. 63), has a very simple form, telling us that for any concepts F and G, the number of Fs = the number of Gs if and only if $F \approx G$, where $F \approx G$ (the Fs and the Gs are in one-to-one correspondence) has its usual second-order explicit definition.

---

[2] See especially *Bg* (sect. 23), and *Gl* (sects. 90–1); see also Frege (1880/1).

Following a terminology established in Boolos (1987), I will often refer to the contextual definition as Hume's principle.[3] I will have more to say about the interpretation of this principle in a moment. But first, I want to call attention to two mathematical facts that will form the background to my discussion: the second-order theory with Hume's principle as its only non-logical axiom is *consistent*,[4] $< 0, 1, \ldots, \omega >$ being the basis of a (second-order) model of the theory. And in the context of second-order logic Peano arithmetic is a definitional extension of this second-order theory; in particular one can derive Peano's second postulate – every number has a successor – from it, as Frege was the first to have noticed, and as Crispin Wright (1983) was the first to have fully appreciated some 100 years later. Frege's proof, more or less complete, even by contemporary standards of rigor, occupies sects. 74–83 of *Grundlagen*. Frege's theorem is just the theorem that, in the context of second-order logic, Hume's principle implies that every natural number has a successor.[5]

The philosophical interest of Frege's theorem derives from the thesis that Hume's principle expresses the pre-analytic meaning of assertions of numerical identity. Frege's basis for this contention was his belief that the equivalence relation of one-to-one correspondence is "conceptually prior" to any notion of number. But it is difficult to find an explicit argument for the conceptual priority of one–one correspondence. *Grundlagen* contains what appears to be an indirect argument for this contention – an argument

---

[3] So named because when introducing the contextual definition Frege quotes the following passage from Hume: "We are possest of a precise standard, by which we can judge of the equality and proportion of two numbers; and according as they correspond or not to that standard, we determine their relations without any possibility of error. When two numbers are so combin'd as that one has always an unite answering to every unite of the other, we pronounce them equal . . ." (Hume (1739, p. 71: Book I, Part III, sect. I, para. 5; quoted by Frege in sect. 63 of *Grundlagen*). Hume's terminology is hardly transparent. John Corcoran has suggested that Hume used 'unite' in roughly our sense of *element*; and by "combining" (comparing) two numbers, Hume meant what we would today call the comparison of two *sets*; to "pronounce them equal" is therefore to pronounce the two sets equal in size (rather than to assert their *identity*). This interpretation may remove some of the irony of calling so fundamental a principle of *Grundlagen* 'Hume's principle,' since it would show that Hume was not committed to the "featureless units" theory of numbers attacked by Frege (*Gl*, sects. 29–39). However, Hume's use of 'number' in the sense of 'plurality' is not an exact match to our use of 'set' since, at the very least, it misses the empty set and singletons, and is therefore subject to an objection exactly analogous to one raised (in *Gl*, sect. 28) to the thesis that a number, in our sense of 'number,' is a set in the sense of a 'plurality.' The 'Hume' in 'Hume's principle' should, therefore, be taken with a grain of salt.

[4] See Boolos (1987, pp. 216–20) for a discussion and further references; page numbers for Boolos (1987) are to its reprinting in Demopoulos (1995).

[5] The full resources of standard second-order logic are required in order to derive the induction principle. However, there is a sense in which the proof of the first four postulates – asserting the Dedekind infinity of the numbers – requires much less. It suffices that the cardinality operator be defined over the concepts corresponding to the so-called "Kuratowski-finite" subsets of N. I am indebted to John Bell for this observation, now developed in Bell (1999b).

Logicism and its Philosophical Legacy

by analogy – which is given in connection with Frege's discussion of the direction of a line.[6] There Frege claims that we attain the concept of direction only when we have grasped the relation of parallelism: although the direction of l = the direction of m if, and only if, l is parallel to m, our understanding of direction depends on our grasp of parallelism, rather than the other way round, and this is what accounts for the fact that direction can be given an analysis in terms of parallelism. It is clear that Frege intends to extend this consideration to the analysis of 'the number of Fs = the number of Gs' in terms of the existence of a one-to-one correspondence, and that he regards the two cases as similar in all relevant respects.[7] Indeed, since in the projective-geometric tradition of the nineteenth century, a tradition with which Frege was certainly familiar, 'the direction of l' meant *the point at infinity associated with l*, the analogy with numbers appears closer than we might otherwise have imagined.[8]

Nevertheless, the argument suffers from the fact that the analogy is far from perfect. The direction of l is an object of the same "type" as l; but on Frege's theory of concepts and objects, F and the number of Fs are of a different type.[9] The contextual definition of direction has no effect on the cardinality of the domain of a model of projective geometry; but as we will see in greater detail later, this is emphatically not the case for the contextual definition of number and second-order logic. Finally, although the conceptual priority of the notion of parallelism may seem plausible within the context of the Kantian theory of our knowledge of space to which Frege appeals,[10] it is completely unclear how to generalize from what is plausible in this case to the case of our knowledge of number. Clearly, therefore, if the claim of the conceptual priority of one–one correspondence over number depends on such an argument by analogy, a great deal more needs to be said before it can be seen to be compelling. Perhaps the analogy with direction is best viewed as heuristic rather than suasive. But even as a heuristic analogy, I think it is potentially misleading in its suggestion that the notion of conceptual priority on which it depends is clear independently of the Kantian framework of Frege's discussion.

I think that there is implicit in *Grundlagen* a more promising line of defense of the centrality of one–one correspondence to our understanding

---

[6] The argument is developed in *Gl* (sects. 64–7).   [7] See, for example, *Gl* (sect. 65, footnote 1).

[8] See Wilson (1992) for an illuminating discussion of the geometric context of Frege's discussion.

[9] And this difference is of great importance to Frege; cf. also the third of the three methodological principles enunciated in the introduction to *Grundlagen*: "never ... lose sight of the distinction between concept and object."

[10] Cf. esp. *Gl*, sect. 64.

of number – one that rests on the "fundamental thought" of sect. 46: namely, the thesis that the content of an ascription of number consists in the predication of something of a concept.[11] This thesis is explicitly and compellingly argued for over the space of many sections of the book and on the basis of a variety of different considerations. If we accept the idea of the fundamental thought, then numbers are known as the numbers of concepts. It is then only a very short step to the conclusion that the content of a statement of numerical identity consists in predicating a relation of concepts. All that is left open is the matter of establishing the correct choice of relation whose predication is involved in statements of the form: 'the number of Fs = the number of Gs.' The suggestion that such identities should be understood to assert sameness of number is hardly informative; but the proposal that they should be understood in terms of one–one correspondence is an informative and natural extension of the fundamental thought.[12] So if we grant the need for an analysis of statements of number at all – if, in other words, we grant that statements of number are not already "perfectly in order" – then, relative to the correctness of the contention of the fundamental thought, Hume's principle is almost forced as a central component of any analysis of 'the number of Fs = the number of Gs.' The principle then suggests itself as a condition of adequacy which any account of number must satisfy,[13] a characterization of the principle that comports well with Frege's practice: once Frege's explicit definition of a cardinal number as a class[14] of equinumerous concepts has been shown to imply Hume's principle, this definition is no longer appealed to, and so far as the mathematical development of the theory of natural numbers is concerned, the account given in *Grundlagen* proceeds from just this principle.

Frege's proof of Peano's second postulate establishes the existence of an appropriate family of representative concepts: with the exception of the first, each representative concept is characterized as holding of precisely the cardinal numbers less than its number: 0 is the number of the concept under which nothing falls, 1 is the number of the concept $x = 0$, and, in general, $n + 1$ is the number of the concept under which the first n numbers fall. Although Frege's proof does not require that we do so, it is interesting

---

[11] Here I follow Michael Dummett's suggested translation rather than Austin's; see Dummett (1991a, p. 88).

[12] This point is also argued in Frege's correspondence with Husserl, collected in McGuinness (1980).

[13] As Dummett has observed; cf. Dummett (1991a, p. 201).

[14] Classes (extensions of concepts) are introduced in *Gl* (sect. 68).

to observe that if we pass to the sequence of extensions of such concepts, we have, in *Grundlagen*, a clear anticipation of the construction of the finite von Neumann ordinals.

More explicitly, Frege's proof of the infinity of the numbers proceeds by defining an appropriate sequence of representative concepts:

N_0: $[x: x \neq x]$
N_1: $[x: x = $ the number of the concept $N_0]$
N_2: $[x: x = $ the number of N_0 *or* $x = $ the number of $N_1]$
etc.

The existence of the numbers of each of the concepts N_0, N_1, ... is established using only logically definable concepts and those objects whose existence can be proved on their basis using Hume's principle. Nowhere in this construction is it necessary to appeal to extensions of concepts.

It is, however, possible to make use of a consistent fragment of the theory of extensions,[15] and when we do so, the comparison with later developments, alluded to above, becomes quite obvious. To see this, suppose we limit the introduction of extensions to those belonging to the concepts belonging to the sequence $< N_0, N_1, ... >$. This yields a new sequence of concepts:

N_0: $[x: x \neq x]$
E_1: $[x: x = $ the extension of the concept $N_0]$
E_2: $[x: x = $ the extension of N_0 *or* $x = $ the extension of $E_1]$
etc.

whose extensions:

$\emptyset$
$\{\emptyset\}$
$\{\emptyset, \{\emptyset\}\}$
etc.

are the finite von Neumann ordinals. Summarizing: without the fragment of the theory of extensions, the numbers are the "objects"

$$0, 1, 2, ...$$

[15] Basically, we require only that fragment which guarantees the existence of extensions for "numerical concepts." In this quasi set-theoretical context, Hume's principle becomes a kind of Axiom of Infinity. See Demopoulos and Bell (1993), and Bell (1995).

associated with the sequence of concepts

$$N_0, \; N_1, \; N_2, \; \ldots.$$

By including the consistent fragment of the theory of extensions it is possible to identify the numbers as the sequence of extensions of the concepts

$$N_0, \; E_1, \; E_2, \; \ldots;$$

and these are just the finite von Neumann ordinals. Judged from the standpoint of their role in the key mathematical argument of the book, it has been suggested that this family of extensions (rather than the family of equivalence classes of equinumerous concepts) form the true Frege finite cardinals.[16] I will return to this suggestion in Section 9.5, below. I hope these brief expository remarks suffice to indicate how, in its mathematical development, Frege's theory provides an elegant reconstruction of our understanding of the cardinal numbers, one that can be carried out quite independently of the portion of his system which led to inconsistency.

Before concluding this part of my exposition, there is another comparison with a later development worth making, one concerning a subtlety that arises in connection with the interpretation of the logical form of Frege's cardinality operator. To see it, consider the representation of Frege's system of concepts and objects from the perspective of simple type theory: Fregean objects are represented by *individuals*, first-level concepts by *sets* (of individuals), and second-level concepts by *classes* (of sets); most importantly, a Fregean number is represented by an individual, and hence by an entity of lower type than the set to which it is assigned. Thus, for Frege, Hume's principle involves what we might call a type-*reducing* or type-*lowering* correspondence, since the cardinality operator takes us from a first-level concept to an object; that is, from a set to an individual. By contrast, for Whitehead and Russell, a cardinal number is a class, so that for them the cardinality operator involves a type-*raising* correspondence. The difference is important, since Frege's proof of the infinity of the numbers depends essentially on the type-reducing character of the cardinality operator: it is what ensures that Hume's principle will have only infinite models. If the operator is type-raising, it is possible to have models of such a transformed "Hume's principle" in which the cardinality of the collection of classes – and therefore the cardinality of the numbers – is finite. Indeed, for Whitehead and Russell the cardinality of the collection of classes is bounded

---

[16] A point that is urged in Boolos (1987, pp. 227 and 229).

by the cardinality of the collection of individuals and, as is well known, to ensure the infinity of the numbers, it is necessary to assume the infinity of the collection of individuals. What is perhaps less well known is the fact that it is a central theorem of *Principia* (vol. II, *124.57) that the ordinary infinity or "noninductiveness" of the collection of individuals suffices – without the assumption of the Axiom of Choice – to prove the *Dedekind* infinity of the Russell–Whitehead cardinals. As noted in Boolos (1994), this is Russell's analogue of Frege's theorem.*

9.3

The context principle was first formulated in the introduction to *Grundlagen* as one of three "fundamental principles" guiding the enquiry: "never . . . ask for the meaning of a word in isolation, but only in the context of a proposition." The exact significance of the principle in Frege's development of his account of the numbers is not entirely straightforward; nor is its relation to the contextual definition particularly simple. If the context principle is interpreted as a principle governing reference, it suggests that reference to abstract objects, and specifically to mathematical objects, can be achieved once we have established the truth of certain key propositions into which they enter.

The relevant contrast is with a "naive Platonism" – this is not a historical claim, hence the scare quotes – which would suppose that we can refer to numbers independently of our knowledge of any such truths. As I intend it, naive Platonism is the view that knowledge of reference "precedes" knowledge of truth. This happens in two ways: first, for naive Platonism, the notion of the reference of an expression is not necessarily tied to the contribution it makes to the truth of sentences in which it may occur as a semantically significant component; in other words, on this view, the concept of reference – even reference to abstract objects – is capable of being understood quite independently of any notion of "semantic value." Secondly, knowledge of the particular referent of an expression which purports to refer to an abstract object – rather than our understanding of the concept of reference for such an expression – does not depend on knowledge of truths.

---

* *Note added 2012*: The significance of these differences for *Principia*'s reconstruction of our knowledge of arithmetic and Frege's analysis of arithmetical knowledge is something I came to see only recently. See the discussion in Section 1.3, above.

In sect. 62 of *Grundlagen*, Frege makes it clear that he sees his use of the context principle as a rejection of what I have called the naive Platonist's view of truth and reference in favor of the contention that they must be understood together. So understood, the context principle seems to have persisted through *Grundgesetze*; if in *Grundlagen* the point of the principle is that truth and reference are coequal, in *Grundgesetze*, its point is that *reference to truth* (to the True) – the reference of sentences – is integral to our concept of the reference of all other expressions. But Frege's use of the context principle goes further than the claim that the concepts of truth and reference must be understood together. Frege takes the additional step of suggesting that the reference of numerical singular terms like 'the number of *F*s' is something we achieve with the recognition of the truth of a distinguished class of statements.

For Frege the statements that are accorded a special role in settling questions of reference are what he calls "recognition judgments"; in the case of number words (and other numerical singular terms) recognition judgments involve the relation of identity, so that for this case, recognition judgments are said to require a "criterion of identity" for numbers. Hume's principle provides such a criterion of identity; its instances form a class of statements associated with the numbers which allow us to say when the same number has been "given to us" in two different ways, as the number of one or another concept. In conjunction with the context principle, our knowledge of the truth of such recognition judgments accounts for our access to the numbers. If the contextual definition restricts the cardinality operator to a particular mapping from first-level concepts to objects, then, assuming that the reference of the relevant concept-expression has been fixed, what is required in order to know the reference of a wide class of numerical singular terms will also have been settled. This is why the equivocation one sometimes finds in the secondary literature between *contextual definition of number* and *contextual definition of the cardinality operator* is not so important as it might at first appear: to the extent that the contextual definition successfully "picks out" the cardinality operator, it must also determine what is required by reference to the cardinal numbers.

The combination – context principle plus criterion of identity – forms the forerunner of the notion of *sense* which Frege was to introduce several years later in his mature theory of meaning and reference.[17] The essential idea of that theory seems already foreshadowed in *Grundlagen*'s account of our reference to the numbers. Hume's principle determines the condition

---

[17] This idea is developed in DeVidi (1996).

which must obtain for the truth of a recognition judgment. Given the truth of a recognition judgment, the context principle entitles us to infer that the associated numerical singular terms are referential. Frege's central philosophical application of the context principle thus consists in the claim that knowledge of the truth of Hume's principle and of the statements expressed by recognition judgments suffices for knowledge of reference. If this approach were successful, it would resolve the problem of reference to the numbers without resorting to a naive Platonist picture, according to which numbers and other abstract objects are "ostended in intuition," where intuition may include – but may also be broader than – the notion of intuition which arises in connection with the Kantian tradition and the issue of the autonomy of arithmetic.

If we accept the proposal that the context principle is concerned with the relative priority of truth and reference in both of the senses distinguished earlier, then the problem of securing the reference of numerical singular terms reduces to the plausibility of the account of our knowledge of the truth of the contextual definition. The proposal, that the context principle is concerned with the relative priority of the semantic categories of truth and reference, should therefore be contrasted with Wright's suggestion that "the more controversial aspect of the context principle [is its assertion of] the priority of syntactic over ontological categories."[18] For Wright, the "syntactic priority thesis" is important because it secures the *objectual* reference of numerical singular terms. But the thesis is too weak to support an answer to the central question of *Grundlagen*: 'How, then, are the numbers given to us?' The syntactic priority thesis tells us that the numbers are given to us as objects *provided that they are given to us at all.* That the numbers are given to us *as objects* may plausibly be held to follow from the syntactic character of numerical expressions. But surely the point of Frege's question is to ask after the basis for the fact that the numbers *are* given to us, that numerical expressions *do* refer – whatever the basis for the assumption that they refer to *objects*. An answer to this question is mandated by the inadequacy of traditional answers, all of which have recourse either to ideas (that is, to experience) or to intuitions (principally, geometric intuitions). The burden of any argument that numerical singular terms refer – at least the burden of any argument based on the context principle – must be borne by its account of our knowledge of the truth and adequacy of the contextual definition.

---

[18] Wright (1983, p. 51). The whole of Wright (1983, sect. viii) is worth consulting. See also Wright's especially accessible formulation in Wright (1990, sect. 2). Wright's current position is discussed in the appendix to this chapter.

I believe that the emphasis on the syntactic priority thesis has tended to deflect attention from the primacy of this issue.

9 · 4

Is it plausible that the context principle, in conjunction with the contextual definition, can successfully show how reference to the numbers is achieved? Frege himself apparently thought not. He chose to follow the practice (which became widespread in the nineteenth century) of passing from the criterion of identity, given by an equivalence relation, to the quotient structure of equivalence classes determined by the relation – what has come to be known as definition by (logical) abstraction.[19] The so-called Frege–Russell explicit definition of the cardinal numbers as equivalence classes (under one–one correspondence) of concepts (or, in Russell's case, classes) is a paradigm case of definition by abstraction. The point I have been emphasizing in my exposition is that Frege's theorem and the mathematical development of his theory of the natural numbers does not require the step to equivalence classes – in fact, does not require any notion of class at all. But success on the mathematical front is no guarantee of a successful account of reference to the numbers, let alone reference to abstract objects. Have we come this far only to discover that the elaboration of this philosophical program *does* require the notion of class (Frege's extension of a concept) after all?

In *Grundlagen*, Frege appears to be committed to the view that a solution to the problem of reference to the numbers must carry with it an elimination of every degree of indefiniteness which the pre-analytic notion possesses; and he appears to have proceeded as he did because he held that since the contextual definition is inadequate to the task of determining the truth conditions of *all* cases of numerical identity, it needs to be supplemented by an appeal to the theory of definition by abstraction, with equivalence classes of equinumerous concepts providing the proper referents of numerical singular terms. Frege saw the incompleteness of the contextual definition in very general terms. It is cryptically presented with a question of the form: 'What prevents us from accepting the identity: "The number of Jupiter's moons = Julius Caesar"?'[20] In the secondary literature the problem this question raises has come to be known as "the Julius Caesar

---

[19] The terminology can be confusing: Russell (1903, sect. 210) attributes to Peano a notion of definition by abstraction that corresponds to what I have been calling contextual definition.

[20] See *Gl*, sects. 56, 66, 68, and 107.

problem"; and in *Grundlagen*, it was supposed to have been resolved by the introduction of the explicit definition of the numbers in terms of extensions. But as several commentators have noted, once extensions have been introduced to resolve the Julius Caesar problem for numbers, an analogous problem arises for classes.

If the claim that the contextual definition captures our pre-analytic understanding of the numbers were correct, we could argue that any deficiency in the contextual definition's ability to resolve all questions of numerical identity is attributable to a deficiency already present in the pre-analytic notion.[21] The difficulty is that although the pre-analytic notion and its analysis in terms of Hume's principle are both compatible with making a stipulation which would exclude the identification of Julius Caesar with the number 4, pre-analytically it is immediately evident that Julius Caesar is not a number. The fact that Hume's principle is consistent with the possibility that Julius Caesar is a number therefore shows that it fails to capture the whole of our pre-analytic understanding.

I want to suggest that although the contextual definition fails to give a complete account of our notion of number, it does succeed in the case of our number-theoretic use of numerical singular terms. We should view Frege's Julius Caesar problem as showing how clarity concerning our number-theoretic uses of numerical expressions – even when it succeeds in making evident how the numbers are applied in counting – may fail to yield a full account of everything we take ourselves to know about numbers. Whatever novelty my suggestion contains is centered on the light it casts on the connection between the introduction of extensions and the Julius Caesar problem; i.e. on how it represents Frege's conception of the difference between a successful account of all possible uses of numerical expressions – including the extraordinary uses that occur in connection with the Julius Caesar problem – and one which restricts itself to our pure and applied number-theoretic uses of them.

9.5

It is a commonplace that a satisfactory account of the mathematical use of numerical singular terms does not require that there be a *unique* ω-sequence within which they refer. Our purely mathematical use is exactly as

---

[21] In this connection, Dummett has suggested that the acceptance of such an "incomplete analysis" requires a constructivist interpretation of the existence of mathematical objects, and the rejection of classical logic. The topic is far too large to broach here. See the final chapter of Dummett (1991a).

Dedekind presented it: *any* ω-sequence constitutes a suitable basis for the domain of an interpretation of the theory of the natural numbers. Richard Heck has persuasively argued that at least by the time of *Grundgesetze*, Frege was well aware of this fact – indeed, sufficiently well aware to have proven that his "basic laws of arithmetic" are *categorical*.[22] This replication of Dedekind's celebrated categoricity theorem[23] is borne out by a close examination of Frege's proof of Theorem 263 of the first volume of *Grundgesetze*. I am proposing that if we regard *Grundlagen* with some interpretive charity, then Frege should be understood as having seen that his account of our number-theoretic use of numerical singular terms, given in terms of Hume's principle, was complete in the sense that his basic laws of arithmetic yield a categorical characterization of the numbers and therefore fix the reference of numerical singular terms up to isomorphism. Moreover, the fact that Frege's derivation of his basic laws proceeded from a statement expressing a criterion of identity showed them to yield a categorical characterization of the numbers as *objects*.

But unlike Dedekind, Frege's interest in a theory of number was not exhausted by its ability to cover purely number-theoretic applications. The difference between them is already signaled by Frege's derivation of the basic laws of arithmetic from a principle which explicitly ties the numbers to one–one correspondence, and thus to our use of the numbers in ordinary applications involving counting. And this difference underlies another: since Frege sought accounts of *truth* and *reference* as these apply to the totality of our statements of number – rather than merely to statements of number theory – a statement such as

$$(*) \text{ The number of Jupiter's moons } = 4,$$

is not about the fifth position in an arbitrarily selected ω-sequence, but is about the fifth element of the particular ω-sequence we identify as

$$< 0, \ 1, \ 2, \ \ldots \ >,$$

This sequence has a privileged position among ω-sequences, and for Frege, all other ω-sequences are to be regarded from the point of view of

---

[22] See Heck (1995, pp. 323–7). Frege's axiomatization of arithmetic consists of four basic laws which assert: (1) the successor relation is functional; (2) no number is related to itself by the ancestral of the successor relation; (3) every number has a successor; and (4) the numbers consist precisely of 0 and its descendents (relative to the ancestral of successor) – the analogue of mathematical induction. These are obviously equivalent to the Peano–Dedekind axioms.

[23] Dedekind (1888, Theorem 132).

how they lie imbedded in it. On this view, the fifth position of the
ω-sequence

$$< 0, 2, 4, \ldots >,$$

is simply not occupied by the object (*) purports to be about; and the fact
that, under a suitable definition of successor, the ω-sequence < 0, 2, 4, . . . >
also forms a model of the basic laws of arithmetic, is simply insufficient to
warrant its being taken as a representation of the numbers as we know them.
Indeed, if any ω-sequence sufficed, there could be no objection to employing the sequence of Roman emperors, provided its members were sufficiently numerous.

To maintain these claims, Frege required a characterization of the numbers
that would single out one ω-sequence from all the rest. This is what the
explicit definition in terms of equivalence classes of concepts was supposed to
accomplish. If successful, it would have achieved this – not because there is
greater clarity about the claim that Julius Caesar is not an extension than that
he is not a number, but because the definition in terms of equivalence classes
of concepts characterizes the numbers by reference to their use in ordinary
applications; it therefore fixes the reference of numerical singular terms to the
extent that the reference of any singular term is fixed. By contrast, the numbers
which are shown to exist on the basis of the contextual definition are
characterized purely internally, with respect to their predecessors in the
sequence of numbers. The construction of the numbers which Frege gives
must have this internal character if it is to serve the purposes of his proof; but
the explicit definition in terms of equivalence classes of concepts is needed if
the account of number is to be complete in the sense of subsuming our
applications of the numbers. In short, it is precisely because the explicit
definition purports to comprehend all our uses of number – both pure and
applied – that it can claim to give *the* reference of our numerical singular terms.

Frege can grant that our mathematical use of numerical singular terms is
indifferent to whether any particular object has the property of occupying
the *n*-th position of an arbitrary ω-sequence; and he can grant that this use
neither requires nor achieves reference to any particular object. He can also
accept that, number-theoretically, the choice of ω-sequence is arbitrary.
What differentiates his view is the claim to have captured *the* reference of
numerical singular terms. For Frege the choice between the sequence of
numbers given by their definition as classes of equinumerous concepts and
any other sequence is not at all analogous to the choice between two
internally defined sequences of numbers, such as the sequence of numbers

associated with the concepts N_0, N_1, . . ., and that given by the extensions of the concepts N_0, E_1, ... For Frege both of the latter two sequences serve the purpose of accounting for our *number-theoretic* knowledge, just as, set-theoretically, it is evident that both the finite von Neumann ordinals and the finite Zermelo ordinals provide equally adequate set-theoretic reconstructions of the natural numbers. The definition in terms of equivalence classes of concepts is intended to serve an altogether different and more ambitious aim, one that none of these constructions even purports to address.

The consistency of Hume's principle shows that Frege's contextual definition sufficed to secure reference in the restricted sense of fixing an interpretation of our use of numerical singular terms as the objects of a domain of a model characterized up to isomorphism. It was the project of securing reference to a particular sequence of objects which are held to be *the* natural numbers that required the step to equivalence classes of concepts. Were *Grundlagen* proposing just a theory of number, rather than a theory with this further aim, there would have been no need to introduce extensions. The discovery of the contradiction presented an obstacle to this more ambitious goal. From this perspective, the inconsistency involved in the explicit definition should be viewed as a discovery concerning the degree of comprehensiveness it is possible to achieve in connection with the explicit definition of the numbers as classes, rather than as a discovery whose principal effect was to have compromised the status of the numbers as logical objects.

9.6 †

Another unresolved problem, concerning Frege's explanation of the reference of numerical singular terms, is connected with the question of the nature of the truth of Hume's principle. The "mere" truth of Hume's principle is insufficient for Frege's purposes if this is understood to depend on our having first specified the referents of its constituent expressions. If Frege had proceeded in this way, his account of the reference of numerical expressions in terms of the context principle and the truth of the contextual definition would have been circular. I believe that the thesis that arithmetic is part of

† *Note added in 2012*: I now recognize that the kind of justification of Hume's principle that is pursued in this section is not fundamentally different from Carnap's assimilation of pure mathematics to our linguistic framework, one of the central ideas of his seminal paper, "Empiricism, semantics and ontology." Carnap's proposal is discussed at some length in Chapters 2 and 3, above. The preliminary discussion of Frege in Chapter 1 provides the context for this discussion.

logic was attractive to Frege because it held out the promise of a model for the truth of Hume's principle that would avoid this type of circularity.

The central idea I am proposing may be put somewhat anachronistically in terms of our ordinary model-theoretic analysis, according to which a logical truth is one that is true in all models: true-in-M, for all M. Now, even though this analysis appeals to truth-in-a-model, and therefore assumes the notions of interpretation and reference, in the case of the logical truths this dependence turns out to be inessential, since the holding of a logical truth in some *particular* M does not distinguish it from any *other* structure. By contrast, an ordinary mathematical truth, such as the commutative law of group theory, to take a noncontroversial example, holds precisely in those structures which are commutative groups. Hence, the truth-in-M of a mathematical law, as opposed to a logical law, depends essentially on the reference of at least one of its constituent expressions. Thus, even though the use of the model-theoretic framework to explain the truth-in-M of a statement requires the notion of reference, it allows that the situation is importantly different according to whether we are considering logical or nonlogical truths; briefly put, the logical truths make no *special* demands on reference. This, it seems to me, is how the "topic-neutrality of logical truth" is spelled out in a model-theoretic setting. To put the point another way: the referential demands of the logical truths are reflected in the model-theoretic framework as a *whole*, rather than in any part of it; it is in this sense that the topic-neutrality of the logical truths shows them to have a special status vis-à-vis questions of existential commitment. The logical truths require only the minimal existential commitment that *any* M would fulfill, which, of course, is different from saying that the model-theoretic analysis of logical truth is without any existential commitment at all (it is committed to at least one nonempty universe).

Is there a way of adapting this understanding of the topic-neutrality of logical laws to the case of Hume's principle? Is there, in other words, a modified conceptual setting in which an appropriate analogue of topic-neutrality can be found to hold of Hume's principle? I want to suggest that the topic-neutrality of the truths of logic can be extended to an account of the truth of Hume's principle, even if Hume's principle is not itself a truth of logic.

I grant that it would be gratuitous to restrict a model-theoretic investigation of Hume's principle by simply stipulating that the cardinality operator is a logical constant. And I also concede that the cardinality operator is not a logical constant, so that in so far as there are models in which Hume's principle fails, it is not a logical truth. It nevertheless seems plausible to expect of any model which purports to constitute a representation of our conceptual framework that it contain concepts which allow for the unrestricted application of the cardinality operator. This is a property of our concepts which we would expect to find represented in any candidate model of what we know. The problem of the truth of Hume's principle consists in establishing the correctness of Frege's analysis of number in terms of one–one correspondence so that the principle can be seen to hold in all such models. It would then enjoy a status in the class of structures purporting to represent our conceptual framework which is entirely analogous to that possessed by the truths of logic in the class of all models: just as a logical truth makes no demand on truth and reference which is not already implicit in every other truth, the referential demands of the truth of Hume's principle would be held in common by *every* truth, and, in analogy with the truth of the laws of logic and the model-theoretic framework, its truth would be reflected in our conceptual framework as a whole.

There is a way of putting this argument that does not rely – or at least, does not so obviously rely – on the notion of our conceptual framework: the notions of truth and truth-in-M are clearly different but related notions, even though neither is analyzable in terms of the other. But however different the two notions may be, it is possible to replace truth with truth-in-M if we have delineated the model which the world comprises.[24] Although this is an impossibly difficult and utopian task, it is not utopian to suppose that we might isolate some of the main features of that structure, truth in terms of which is equivalent to truth. For example, we might succeed in isolating what parameters such a model will have, given very general features of our understanding of what would constitute an adequate account of such a model. In particular, any model that might plausibly represent our notion of truth *simpliciter* will contain both "the" cardinal numbers and the concepts associated with them in the development of Frege's proof, outlined in Section 9.2, above.

But then, if Hume's principle yields the correct analysis of the truth of recognition judgments – briefly, of numerical identity – it will be true in any model which is a candidate for representing our notion of truth *simpliciter*; in this sense, Hume's principle is a completely general truth. And to the degree that the fundamental thought can be said to provide an analysis of what Frege calls "statements of number," Hume's principle can be held to be analytic of statements of numerical identity; in this sense, it is a

---

[24] Compare Gupta and Belnap (1993, pp. 23–4).

necessary truth. The sense in which, on this approach, Hume's principle or the contextual definition is "definitional" is that it advances an account of our pre-analytic notion of numerical identity. The contextual definition is not a mere definitional stipulation but, like the fundamental thought, it purports to capture a pre-analytic notion.

Even if we never achieve a complete specification of the model which the world comprises, it would suffice, so far as the vindication of this account of our knowledge of arithmetic is concerned, that we should know of any such model that it is constrained by Hume's principle. For, if Hume's principle does constrain any such model, then we are justified in supposing that it is *true* – not just true in some model or other, but true in that model, truth in which coincides with truth *simpliciter*.

But if Hume's principle is true, we are also justified, by the context principle, in holding to a limited realism concerning the cardinal numbers. The realism to which we are entitled fails to be a complete vindication of Frege's program because Hume's principle captures the numbers only up to isomorphism; it fails, in other words, to solve the Julius Caesar problem as this was represented in Section 9.5, above. There is therefore a sense in which the realism concerning the numbers which these considerations vindicate is attenuated relative to that which attaches to realism regarding the physical world: we do not think of the furniture of the world as specifiable only up to isomorphism, but regard it as something we will have captured uniquely, once we have succeeded in characterizing the model which the physical world comprises. Frege seems also to have partially conceded this qualification when he sought an account of the numbers which would single out one ω-sequence from all the rest by characterizing the numbers in terms that would comprehend their applications. So long as the numbers are captured only "structurally," there is a certain conventionalism which attaches to the assertion of their existence: *any* ω-sequence will serve as the sequence of numbers. This is a feature without an analogue in our conception of the constituents of the physical world. A complete vindication of "Fregean Platonism" would require being able to distinguish a unique such sequence. But for this we require resources that go beyond the contextual definition. It is precisely at this point that the notion of class is required (as we saw in Section 9.4). Nevertheless, we have achieved something more than the conclusion that the contextual definition is merely true-in-a-structure. Instead of this rather weak result, we have explained how it can be understood to be *true*; as such, it has a distinguished status among mathematical principles, as do the referential commitments of its constituent expressions. Though a slight advance, it is an advance nonetheless.

I believe that the foregoing proposal has certain additional features which recommend its consideration. Here I would just like to emphasize that it in no way depends on Frege's metaphor, according to which equivalences like Hume's principle involve the creation of a "new concept by the recarving of a content" (*Gl*, sect. 64); nor does it require the closely related idea that reference to the relation of one–one correspondence of concepts involves a "tacit" reference to numbers – even when that relation is taken as primitive. On the present proposal, the explanation of the significance of Hume's principle is much simpler: the contextual definition is a constraint on the class of models that are capable of adequately representing our conceptual structure and our notion of truth *simpliciter*, a constraint which bears a direct analogy to that imposed by the logical truths on the class of all structures. By Frege's theorem, in order for any such structure to satisfy Hume's principle, it must contain "the" numbers. The existence of the numbers – the "limited Platonism" of this view – as well as our success in referring to them is justified to the degree that we are justified in supposing that the world is modeled by such a structure.

9 · 7

In conclusion, I want to return to the objective, initially formulated in *Begriffsschrift*, of presenting an account of arithmetic of sufficient generality to show its autonomy from geometry and kinematics – the so-called "refutation of Kant." To meet this objective, Frege needed to show that certain basic laws of arithmetic – various structural properties of the ancestral of the successor relation, induction on the natural numbers, and the infinity of the natural number sequence, to take just three prominent examples – are derivable from sufficiently general analyses of the notion of a sequence, the number concept, or our concept of the individual numbers, as the case may be. How did the discovery of the paradoxes of set theory affect this aspect of Frege's program?

Important as they were, I believe that the set-theoretic paradoxes have unnecessarily interfered with our ability to appreciate Frege's achievement. This is not to say that the paradoxes are without effect. For example, it is unclear how to articulate an account of the numbers that will comprehend all our applications of them. More specifically, it is unclear how to achieve this while preserving the thesis that numbers are objects. To the extent that they are objects, they fail to be "comprehensive"; and to the extent that we are in possession of a reconstruction which makes the numbers suitably comprehensive, they fail to be objects, i.e. arguments to first-level concepts. The paradoxes thus appear to preclude a unified theory of number.

But if we ignore the issue of whether Hume's principle is analytic or a principle of logic and concentrate instead on the more tractable question as to whether it yields an analysis of number possessing the requisite *generality*, then it is evident that for its deployment in Frege's theorem Hume's principle *is* appropriately general: after all, it allows for the application of the cardinality operator to any (sortal) concept. And the analysis of induction as well as the definition of the ancestral and the proofs of its properties are all carried out within pure second-order logic – which is to say, they are carried out without any appeal to the notion of the extension of a concept. None of these achievements is affected by the discovery of the paradoxes, and more than a century later they still stand.

The assessment of Frege's success has stalled on the question whether second-order logic is really logic. I believe this also to be unfortunate, since it has obscured one of Frege's principal philosophical achievements, namely that of having shown that reasoning by induction and our knowledge of a Dedekind-infinite system depend on principles which, whether or not they are counted as logical, are appropriately general in their application. In particular, they do not rest on any Kantian notion of intuition of the sort Frege sought to refute. It may be necessary to concede that these demonstrations rest on intuition in a familiar, psychologistic, sense of the term; what has been missed is how little is conceded by such an admission.

APPENDIX

There is a recent exchange between Crispin Wright and Michael Dummett which came to my attention shortly after the body of this paper was completed.[25] This appendix reviews some of the issues raised by this exchange in order to clarify the position advanced here. I have not attempted a complete account of the matters under discussion but have organized my remarks around Dummett's charge of vicious circularity in Frege's and Wright's deployment of the contextual definition and the significance of the so-called "bad company" objections to the contextual definition. Wright and I are agreed that there is something hidden in Frege's project which remains to be uncovered, but we may differ on exactly what it is, on how to extract it and, perhaps, on exactly how far it can take us toward the solution of our problems. Throughout this appendix I have used 'contextual definition' rather than

[25] See Wright (1997, 1998a and 1998b) and Dummett (1998). The exchange between Bob Hale, who supports Wright, and Peter Sullivan and Michael Potter, who oppose him, is concerned with related issues and is also well worth consulting, especially for the reader whose principal interest is the Julius Caesar problem and Wright's approach to it. See Hale (1994) and Sullivan and Potter (1997).

'Hume's principle'; not only is this a more neutral term on which all parties agree, but it also highlights the fact that it is the interpretation of Hume's principle as a kind of "definition" that is at issue.

What is at the basis of Dummett's claim that an unavoidable circularity vitiates Frege's and Wright's use of the contextual definition? As we know, Frege rejected the use of the contextual definition for the purpose of introducing numerical singular terms. Wright wishes to revive this idea. The reason why Dummett can treat their views together is that Frege seems to have pursued a strategy similar to the one rejected in the case of the contextual definition of numerical singular terms when, in *Grundgesetze*, he came to deal with the introduction of singular terms for extensions of concepts. Our focus will be restricted to the elaboration of Frege's strategy in connection with numerical singular terms.

For Wright, the contextual definition is a definition only in an extended sense: although it introduces a class of new terms, and in this respect functions like a definition, it fails to satisfy criteria of eliminability and noncreativeness. It is "impredicative" in the sense that the referents of the terms it introduces fall within the range of the quantifiers of the "defining" condition of its right-hand side. To see this, simply recall that first-order quantifiers occur in the unabbreviated statement of the expression for one–one correspondence and, by the type-reducing character of the cardinality operator, the objects it assigns to concepts – the referents of the terms introduced – must fall within the range of these first-order quantifiers. Notice, however, that in its *Russellian* form, the contextual definition is not at all impredicative and that this is a feature which is peculiar to its *Fregean* formulation.

It is important to emphasize that for Dummett impredicativity is not by itself an objectionable feature of the contextual definition. It is only when the contextual definition is regarded as a suitable vehicle for introducing the referents of numerical singular terms, or equivalently, when it is taken to have circumscribed the domain of arithmetic, that its impredicativity is for him problematic. In particular, Frege's proof that the contextual definition implies the infinity of the numbers cannot be taken to have established, in any intuitively acceptable sense, the *existence* of the numbers, since their existence is already presupposed by the impredicative character of the contextual definition. Dummett allows that once the existence of the numbers has been granted, the proof may legitimately be interpreted as having established their infinity; but this is a much less ambitious contention.

Dummett has a second, related objection that does not concern the interpretation of Frege's theorem as a proof of the existence of the numbers but addresses the feasibility of using the contextual definition and the context principle to explain the reference of numerical singular terms. Basically, the objection begins with the observation that we cannot argue from the truth of the contextual definition to the reference of its constituent terms, since their reference must be settled before its truth can be evaluated: to address the truth of the contextual definition we must be in possession of an independent specification of the domain over which it is supposed to hold. Frege could have avoided the demand for such a specification, had it been possible to establish the truth of the contextual definition by deriving it from his basic laws and a satisfactory account of the reference of expressions for the extensions of concepts. But since it is the essence of Wright's approach to *begin* with the contextual definition itself, we need to be given an account of its truth. However, it is unclear how to give such an account without presupposing a solution to the problem of the reference of numerical singular terms – which is, of course, the very problem to which the truth of the contextual definition was supposed to provide the solution.

A major burden of the argument of Wright (1997 and 1998a) is to show that the contextual definition can serve to "introduce us" to the cardinal numbers without appealing to our prior understanding of numerical singular terms. That it can do so is, for Wright, a consequence of a general theory of how abstract singular terms are introduced. In this respect Wright's position differs from Frege's, since it is clear that Frege uses quite diverse theories of knowledge to motivate his understanding of the epistemic priority of the right- over the left-hand sides of one or another contextual definition. For example, in the case of *direction*, Frege's argument rests, not on a general fact about contextual definitions, but on the Kantian theory of the primacy of intuition in our knowledge of geometry. However, Wright's aim is to expose a general role for contextual definition in an explanation of the cardinality operator. For Frege's contextual definition to successfully fulfill this role, Wright argues that it suffices to show that someone could follow the use of the cardinality operator in Frege's proof without having made any explicit assumptions about the existence of the numbers.

Whether we use the contextual definition to introduce the numbers or to establish their existence, it will be necessary to address Dummett's charge of circularity. This is what Wright attempts to do in connection with the thesis that it is possible for someone to attain concepts of the individual numbers by first laying down the contextual definition as a *stipulation* governing numerical singular terms, and then proceeding with the elaboration of the consequences of this stipulation as these unfold through the various steps of

Frege's proof of his theorem. On Wright's account, the contextual definition is an "abstraction principle," a stipulation governing the use of the new
class of abstract singular terms it introduces, rather than the expression of an
analysis of a preexisting notion.

Suppose we grant that the numbers can be explained to someone in the
manner described by Wright. Does it follow from the fact that someone has
such an understanding of numerical singular terms that he has succeeded in
referring to the numbers? Can we identify the problem of showing how
reference to the numbers is achieved with showing how, using the stipulation embodied in the contextual definition, we can come to use a wide
enough class of numerical singular terms to enable us to reconstruct the
proof of Frege's theorem? Although Wright may be able to claim that his
account requires fewer demands on the contextual definition than does the
position criticized by Dummett, is it clear that the conclusion which it
supports is not commensurately weaker?

Leaving these questions aside, the main difficulty, as I see it, is that an
explanation of the truth of the contextual definition and the reference of
numerical singular terms is incomplete if it is not able to tell us "in
advance" how things will turn out for the contextual definition. If
there is nothing in the account to *force* a positive decision on the truth
of the contextual definition, then something more needs to be said;
otherwise we will be left in a position like Russell's with the Axiom of
Infinity. I believe that this is indeed the situation with Wright's account
of the contextual definition, a point that can perhaps be put more clearly
if we turn our attention to the so-called "bad company objections"
to Wright's assimilation of the contextual definition to abstraction
principles.

Boolos (1990) introduced a number of sentences with very different model-
theoretic properties from one another but with an equal claim to being treated
as abstraction principles that introduce a class of abstract singular terms.
Wright (1998b) considers a modification of one of Boolos's examples in
order to motivate a constraint on "admissible" abstraction principles.
Wright's example involves a type-reducing operator which takes two concepts
to the same object just in case their symmetric difference is finite:

NP: $n(F) = n(G)$ if and only if $\{x : x$ is F but not G or G but not $F\}$ is finite.

Wright shows that NP holds in finite domains but fails if the domain of
individuals is infinite and the range of the concept variables is the full power
set of this domain.

The restriction to the full power set is necessary, since there is a Henkin model with an infinite domain in which NP is true; in this model the domain of the concept variables is also countable: let the domain N of the Henkin model be the set of natural numbers, and put $S(N) = \{F: F$ is a finite or cofinite subset of $N\}$ (here identifying a concept with its extension). Observe, by the way, that $S(N)$ is a field of sets just like $P(N)$ – the full power set of N. Now define $n(F)$ by the condition that $n(F) = 0$ if F is finite and $n(F) = 1$ if F is cofinite. Clearly the equivalence relation of NP – F and G are equivalent if their symmetric difference is finite – makes F and G equivalent if and only if F and G are both finite or both cofinite, so they will be assigned different values by n only when one is finite and the other is cofinite. (Proof: Clearly, if F and G are both finite or cofinite, F and G are equivalent, since in both cases we are dealing with a union of intersections, each of which is finite. For the contrapositive, take F finite and G cofinite. Then the complement of F is cofinite. But the intersection of two cofinite sets is cofinite, and the union of a cofinite set with *any* set is infinite. Hence F and G are inequivalent, and thus must be assigned distinct objects.)

To see the difficulty that such a "bad company" example poses, assume (for whatever reason) that we are committed to a standard interpretation of second-order logic. Then, as Wright has shown, were we to adopt the stipulation embodied in NP we would be restricted to models having only finitely many objects. *Since the contextual definition holds only in infinite domains, our adoption of NP would preclude us from stipulating that numerical singular terms should be used in accordance with the contextual definition.* But if we take seriously the idea that abstraction principles are merely stipulations governing the use of the singular terms they introduce – i.e. that they are conventions which we freely lay down – we might easily defeat the truth of the contextual definition by an "incorrect" initial choice of abstraction principle. This would occur if, for example, we had initially chosen the stipulation embodied in NP, rather than the contextual definition. But if abstraction principles are stipulations, and thus the contextual definition is just one stipulation among many, how can we make sense of the idea that there is a *right* initial choice?

Notice that this objection does not apply to a view which takes the contextual definition to be responsible to a pre-analytic conception of the numbers. On such a view, the account of NP would be that it is true in some structures and false in others. This of course is not to argue that NP is analytically false for having restricted the domain of objects to a finite cardinality. The point is rather that while the contextual definition requires the infinity of the *numbers*, it is indifferent to whether the *general notion of an object* should or should not have infinitely many instances in alternative domains; it is, therefore, not really a competitor with an abstraction principle like NP. Indeed, if it is constitutive of an abstraction principle that it is a stipulation, then on the view advanced here, the contextual definition is simply not an abstraction principle.

As I indicated at the very beginning of our discussion of bad company examples, Wright is not without an answer to the difficulty posed by NP which he uses to motivate a constraint on abstraction principles. Wright's suggestion is that an abstraction principle is *admissible* provided that it is compatible with the existence of a countably infinite domain. But now, if we take this route, how far have we really traveled beyond Russell's postulation of a noninductive set of nonlogical objects?

Let me close with the formulation of a general problem whose solution might provide the needed motivation for Wright's approach. Because of its connection with Basic Law v, let us call it *Frege's problem*: to determine those equivalence relations and subsets of a countably infinite domain over which they are defined such that the quotient of the family by the relation is injectable into the domain. The Henkin model for NP described earlier, together with the equivalence relation of NP, are a solution for one such case, and the standard model (full power set) and one–one correspondence are a solution for another; but what is the theory that ties them all together? Indeed, *is there*, in any interesting sense, a theory which ties them all together? If there is such a theory, it might very well dispel any inclination to see Wright's admissibility constraint as a piece of theft rather than honest toil.

ACKNOWLEDGMENTS

Support from the Social Sciences and Humanities Research Council of Canada is gratefully acknowledged. It was a pleasure to present the paper on which this chapter is based to the Buffalo Logic Colloquium and to the Philosophy Department colloquia of the University of Maryland, the University of California at Davis, the University of Toronto, and the University of St. Andrews. I wish to thank Crispin Wright and Michael Dummett for sharing pre-prints of their work with me and for discussions and correspondence on the topics dealt with in the appendix; thanks also to John Bell, Peter Clark, and Gerry Callaghan for discussions on various issues. I am particularly grateful to Anil Gupta for his perceptive comments on earlier drafts of this paper and its appendix.

CHAPTER 10

The *1910* Principia's *theory of functions and classes*

It is generally acknowledged that the first edition of Whitehead and Russell's *Principia Mathematica* – the 1910 *Principia* as I shall call it – does not deny the existence of classes, but claims only that the theory it advances can be developed so that its commitment to classes is readily seen to be merely apparent. This *ontological* application of "the method of contextual analysis" is widely viewed as the central philosophical contribution of Russell's theory of descriptions to the formulation and defense of logicism; and *Principia*'s "no-classes theory of classes" is regarded as one of the method's most striking achievements. In the discussion that follows, I present a reconstruction of *Principia*'s theory of functions and classes that is based on Russell's *epistemological* applications of the method of contextual analysis. The reconstruction I propose is not eliminativist – indeed, it explicitly *assumes* the existence of classes – and it possesses certain advantages over the no-classes theory advocated by Whitehead and Russell – advantages that are concentrated on the light it casts on the reasons in favor of ramification.

My discussion is organized as follows. After reviewing the desiderata a successful reconstruction of the 1910 *Principia* should satisfy, I formulate four "dependency theses" and argue for the primacy of one of them for a logical theory of classes and functions. I then isolate two constraints – which I call "Argument Independence" and "Function Dependence" – that are characteristic of *Principia*'s theory of propositional functions, and I show that the satisfaction of these constraints is forced by Russell's analysis of the paradoxes in terms of a number of principles of "vicious circularity." My main claim is that the vicious circle principles, and, a fortiori, *Principia*'s

Based on my paper "The 1910 *Principia*'s theory of functions and classes and the theory of descriptions," *European Journal of Analytic Philosophy* 3 (2007): 75–93. In addition to numerous additions and stylistic changes, the chapter contains an extended appendix that was not part of the original publication.

notion of a propositional function, are susceptible to a much stronger justification when the 1910 theory – which I take to include the Axioms of Reducibility – is understood along the lines of the epistemological reconstruction I develop. In the appendix I expound Russell's propositional paradox. This paradox provided the initial motivation for a theory of types based on ramification, and it has an evident bearing on Frege's theory of thoughts. I show that the propositional paradox also has important consequences for Dedekind's account of our knowledge of number.

10.1 THE STATUS OF CLASSES

The 1910 *Principia*'s theory of propositional functions and classes is officially a "no-classes theory of classes," a theory according to which classes are superfluous. But it is clear from *Principia*'s solution to the class paradoxes that although the theory it advances holds that classes are superfluous, it does not deny their existence. Whitehead and Russell argue from the supposition that classes *involve* or *presuppose* propositional functions to the conclusion that the paradoxical classes are excluded by the nature of such functions. This supposition rests on the representation of classes by class abstracts and the technique of contextual analysis. But at no stage does *Principia*'s solution to the paradoxes appeal to anything so strong as the denial of classes. Indeed, its account of the paradoxes is a striking vindication of the traditional logical conception of class, since it shows the resolution of the difficulties to depend on the close association between properties (or propositional functions) and classes that is the hallmark of that conception.

It is therefore clear that the status of classes in *Principia* is a matter of some subtlety: any apparent commitment to them is eliminable by the method of contextual analysis, but the "reduction" of classes to propositional functions which such an analysis effects does not, by itself, militate against assuming their existence any more than the reduction of propositions about Sir Walter Scott to propositions that do not contain him as a constituent militates against assuming his existence. Nor would the elimination of classes solve the paradoxes. The key step in the solution of the class paradoxes is the recognition of the "dependence" of classes on propositional functions. Hence whether or not classes are regarded as superfluous, it remains to show that there is a paradox-free notion of *propositional function*. For this and other reasons (to be discussed below) the approach to *Principia* I advocate is one that sets entirely to one side the program of avoiding a commitment to classes or viewing them as superfluous; instead it reconstructs the work within a framework that admits the existence of classes along with propositional functions.

With the exception of *Principia*'s eliminativist sympathies, the reconstruction I am proposing leaves intact almost all of the work's distinctive features while showing it to contain a theory of our knowledge of classes of considerable elegance. Here and elsewhere, when I speak of our knowledge of classes, this should be understood in terms of Russell's distinction between *knowledge of things* and *knowledge of truths*.[1] My idea is to explain what I take to be *Principia*'s implicit theory of our knowledge of classes as a species of our knowledge of things. The theory that emerges bears some similarity to Russell's theory of knowledge by acquaintance and by description, but it is not a straightforward transcription of that theory from the case of things to that of classes. The important similarity to bear in mind is that the theory is first and foremost a theory of how classes can coherently be taken to be the subjects of propositions that are asserted of them. I hope to show that the theory is of interest even if it fails to address everything one might reasonably desire of a theory of knowledge of classes.

There are at least two desiderata that it would be reasonable to demand of a reconstruction of *Principia*'s theory of functions and classes:

(1) The reconstruction should motivate ramifying the simple type-theoretical hierarchy of propositional functions.

To address (1) a reconstruction must show that propositional functions are plausibly represented as satisfying the requirements expressed by ramification. This demand would be satisfied if it could be shown that the relevant vicious circle principles which constitute Russell's diagnosis of the paradoxes are analytic of the concept of a propositional function; establishing this conclusion is a primary aim of the present chapter.

(2) The reconstruction should explain why the Axiom of Reducibility is intrinsically plausible within the framework of the 1910 *Principia*'s theory of functions and classes.

The satisfaction of this desideratum has proven especially elusive. It is an unfortunate consequence of the influence of Ramsey (1925b) that the usual view of the ramified theory of types is that while it may be essential to Russell's solution to the semantic paradoxes, it is at best unmotivated in the case of the theory of classes. And since the theory is too weak to recover arithmetic without an Axiom of Reducibility,[2] extending the theory by the inclusion of such an axiom has come to be regarded as irremediably ad hoc. The present reconstruction seeks to make the ramified theory with reducibility plausible as a theory of our knowledge of functions and classes,

---

[1] Developed in Russell (1912, pp. 44–5 and Chapter 5).   [2] As shown in Myhill (1974).

whatever its usefulness and plausibility as an account of the semantic paradoxes.[3]

The following overview may provide a useful orientation to the reconstruction I am proposing even though it assumes familiarity with notions that will be explained more fully in the sequel.

On my view, the interest of Russell's logicism is almost wholly epistemological: the central point of the ramified theory of functions and classes is to show how, from the assumption that there are certain classes, it is possible to provide an account of our knowledge of them. In particular, we would like to explain how we can have knowledge of the class *N* of natural numbers. *Principia*'s approach to our knowledge of *N* appeals to the method of logical construction, an approach that follows the classical Frege–Dedekind definition of *N* in terms of the satisfaction of all inductive functions of zero. However, Frege's and Dedekind's methodology elides distinctions of *order*. When the definition is reformulated within the framework of ramified types, its success depends on the presence of a sufficiently rich supply of *predicative* functions. By the Axiom of Reducibility there is a plethora of such functions, and as has long been recognized, reducibility is unexceptionable if one assumes the existence of classes. When the Occamism of *Principia* is set to one side and classes are admitted, the ramified theory of functions and classes can readily be seen to contain a compelling account of how we can have knowledge of classes which, like *N*, occupy a central position in the philosophy of mathematics.

The role of reducibility is not dissimilar from that of *Principia*'s Axiom of Infinity. Arithmetic poses two basically epistemological questions for logicism:

(a) What does our knowledge of the Dedekind infinity of the numbers rest upon?

(b) How is reasoning by induction justified?

Principia's answers to both questions depend on the provision of logical constructions. But the constructions succeed only if they are provided with

---

[3] I have throughout been guided by Gödel's fundamental paper (1944), and by Warren Goldfarb's (1989) response to it. I have also benefited from Leonard Linsky's paper (1987), which I have found in many ways insightful. Here I have in mind especially his remark that "[*Principia* presents] an iterative concept of functions, which is formally similar to Cantor's original iterative concept of sets" (p. 37). But I also have a number of difficulties with Linsky's paper: the discussion of propositional functions (on pp. 35–7) relies too heavily on the 1910 introduction's not particularly clear notion of an "ambiguity" to yield a proper reconstruction of the relevant remarks; Russell's emphasis on the extensionality of classes in his letter to Frege (of July 24, 1902 which Linsky quotes on p. 28) hardly shows him to have explicitly formulated "the mathematical or iterative concept of sets" as Linsky implies; and, contrary to Linsky's claim on p. 26, although Frege may have doubted the suitability of Law v as an axiom, until Russell's famous letter he never questioned its consistency.

a sufficiently rich base on which to operate. In connection with (a), there must be infinitely many objects *i* of lowest type for the construction of the numbers to return their Dedekind infinity when they are represented as objects of (simple) type ((i)), i.e. as classes of classes of individuals. In the case of (b) there must be enough predicative functions of the reconstructed numbers to ensure that the class defined by the ramified form of the Frege–Dedekind definition is one for which mathematical induction is recoverable in its full generality. Neither existence assumption is one that *logic* is capable of securing, but if the assumptions they express are correct, *Principia*'s logical constructions succeed in addressing both questions (a) and (b). Hence, far from signifying the failure of the program of a logical foundation for arithmetic, the necessity of such existence assumptions simply confirms the idea that although logic is capable of addressing what is consequent upon various existence assumptions, it cannot settle the existence questions themselves. *Principia*'s success in connection with the epistemology of arithmetic is therefore a kind of triumph for logicism, although in so far as it is forced to give up any claim to having established the apriority of these two existence claims, it is not the triumph a traditional logicist like Frege or Dedekind envisaged.

10.2 TYPES AND ORDERS

The *simple* type hierarchy of propositional functions begins with the type of individuals and proceeds to functions of individuals, functions of functions of individuals, etc. Ramification imposes a system of *orders* on the functions of any type. We will assume that the simple hierarchy of functions is stratified, but will allow orders to be cumulative or stratified. The discussion assumes very little by way of the technical elaboration of the ramified theory and is intended to be broad enough to accommodate any reasonable formalization of ramification which embodies the following four constraints on the assignment of (finite) orders: (i) individuals have order o; (ii) functions have order at least 1; (iii) only propositional functions of an order lower than a function can occur as arguments to the function; (iv) a variable bound by a quantifier occurring in a function can range only over functions of an order lower than the order of the function itself.

The base of the hierarchy can be expanded to include propositions. Provision must then be made for orders of propositions (by contrast with individuals, among which there are no distinctions of order), and the type of propositions must be distinguished from the type of individuals. The hierarchy then continues with propositional functions whose arguments include propositions (and perhaps individuals), functions of such propositions, etc., together with an appropriate system of orders at every level of the expanded hierarchy. Such an expansion is discussed in the appendix to this chapter. My present concern is with the theory of functions and classes involving only the hierarchy of functions based on the type of individuals. I will also ignore the possibility (which *Principia* does not pursue) of extending the hierarchy of types into the transfinite, a possibility Gödel exploited in his investigation of the consistency of the continuum hypothesis.

Intuitively, the order of a propositional function is a rough measure of the complexity of its quantificational structure. Functions of least complexity involve no quantifications over other functions within their type. Such functions are the functions of order 1, and in *Principia* they are called "predicative functions"; those that are not predicative are called "nonpredicative functions." (It should be noted that this terminology is peculiar to *Principia*. On this use of 'predicative,' a predicative function is *not* one whose definition is predicative rather than impredicative, since even a function which occurs primitively and without definition can be predicative in the intended sense.) The predicative and nonpredicative functions form a hierarchy of *predicable* entities based on the type of individuals. The predicative functions form a substructure of this hierarchy which I will sometimes refer to as *the predicative hierarchy*. The members of the predicative hierarchy are logically transparent; the nonpredicative functions are logically complex in the sense that their canonical expression in the language of *Principia* has a quantificational structure that includes quantification over other functions.

By contrast with the hierarchy of predicable entities, classes and individuals form a separate hierarchy of *objects*, one that is simple, not ramified; this hierarchy begins with individuals and proceeds to classes of individuals, classes of classes of individuals, etc. There is no notion of the order of a class, or equivalently, all classes are of the same order.

Principia replaces the traditional logicist's concepts with propositional functions which are a kind of abstract intensional entity, not to be identified with the open sentences of a language of fixed vocabulary or even the indefinitely extendable one of *Principia*. Consequently, the hierarchy they comprise may be as large as any hierarchy of classes whose membership they are intended to determine. This is in keeping with the admission (*Principia*, vol. ii, p. vii) that there are always more propositional functions than there are individuals.[4]

---

[4] There is a significant secondary literature which seeks to establish a nominalistic interpretation of propositional functions as linguistic expressions. There may be an interesting reconstruction of the

The notion of type is readily extended to variables; the type of variables for individuals differs from that of functions of individuals, and both in turn differ from that for functions of functions of individuals, etc. In *Principia*, there are special symbols for variables for functions of lowest order, but no other orders are explicitly indicated by the symbolism. Because of the required division of variables into types and orders, variables are not universal and are not unrestricted in their scope. The loss of the "unrestricted variable" is addressed at the level of the language of *Principia* by the device of typical ambiguity, according to which the type of a syntactic expression for a variable is ambiguous: the same symbol may stand for a variable of any type; and within any type, the same symbol can stand for a predicative function of that type. It is therefore ambiguous what proposition or propositional function is expressed by a formula. Since what matters are only *relative* types, the ambiguity as to type can never lead to a violation of the type restrictions.

The device of typical ambiguity yields a considerable notational simplification. But its philosophical point is that it allows Russell to recover a surrogate for the unrestricted variable of his pre type-theoretic view. As we have just seen, this arises from a combination of two ideas: the variables constituent in a proposition or propositional function are of maximal scope within a type, and the symbols by which variables are expressed are ambiguously interpretable over arbitrary types.

10.3 *PRINCIPIA*'S MODEL OF OUR KNOWLEDGE OF CLASSES

The classical epistemological problem posed by classes can arise in connection with our knowledge of any infinite class. As such it is importantly different from – and clearly epistemologically prior to – the problem of determining the composition of the set-theoretic universe and from the related concern with the independence questions of set theory. The problem is to explain how classes are "given" to us, a problem that has an obvious parallel with Frege's question, in sect. 62 of *Grundlagen*: 'How, then, are numbers given to us, if we cannot have ideas or intuitions of them?' As with numbers, for classes to be acceptable to a logicist it must be possible to

theory of types along these lines, but it cannot be advanced, as it sometimes is, as a correct representation of Russell's intentions in the period under study. Not only is there the point made in the text regarding the cardinality of propositional functions relative to that of individuals, but there are explicit statements to the contrary which simply cannot be passed off as mere slips. Consider, for example, the following passage from Russell (1908, sect. v, p. 80): "A propositional function of x may, as we have seen, be of any order; hence any statement about 'all properties of x' is meaningless. (A 'property of x' is the same thing as a 'propositional function which holds of x.')"

provide an account of our knowledge of them that does not rest on ideas or intuitions. The traditional logicist answer is that classes are given to us as the extensions of concepts.[5] This epistemological question and *Principia*'s nuanced answer to it are our central focus. As we will see in greater detail, *Principia*'s implicit theory of our knowledge of classes is developed within a framework of existence assumptions about propositional functions. The principal justification for these assumptions is that they support the account of *how* certain classes which we unquestionably know *are* known.

The general shape of *Principia*'s theory runs as follows. Knowledge of a class is mediated by knowledge of a propositional function which determines it. Knowledge of various particular classes by means of knowledge of the predicative functions that determine them is assumed to be unproblematic, but it constitutes the exception not the rule. The theory of propositional functions which evolved out of Frege's logical investigations allows for a rich variety of forms of propositional function. The general epistemic situation for which an account is especially pressing is the case in which our knowledge of a class is mediated by knowledge of a *non*predicative function, i.e. by a logical construction based on a limited number of known predicative functions. An account of such cases is especially desirable when the nonpredicative function has an independent foundational interest of its own, as is the case with the Frege–Dedekind definition of the class of natural numbers. According to this definition, the class of natural numbers is the class of all *u* which satisfy every inductive function of zero, where a function is *inductive* if whenever it is possessed by *u* it is possessed by *u*'s immediate successor. Assuming knowledge of *zero* and *immediate succession*, the definition explains our epistemic access to the class of natural numbers in terms of our grasp of a propositional function whose additional complexity is wholly logical; for, under the hypothesis that *zero* and *immediate succession* are understood, *natural number* is explained in terms of the notion, *every propositional function*. The difficulty which the Frege–Dedekind definition presents is one of preserving its success in determining exactly the class of natural numbers while respecting the theory of types and orders.

Taking an uncritically realist view of propositional functions and classes clarifies the special epistemic role functions play in the theory by marking a sharp separation between two types of dependency that classes have on

---

[5] The appeal of this answer was not confined to logicists. It was, for example, strongly endorsed by Hermann Weyl in his book on the continuum. For a discussion, together with relevant references and citations, see Parsons (2002, sect. 1, esp. pp. 380–1).

propositional functions. Of the two dependency theses, the first is the *semantic dependency thesis*:

All truths concerning classes are reducible to the use of class abstracts.

In so far as an abstract employs a propositional function that holds precisely of the members of the class, the thesis implies that classes presuppose their defining functions. And since a class abstract is a definite description, one of the form, 'the class of all *u* such that φ*u*,' the semantic dependency thesis is sometimes expressed by the claim that classes are "incomplete symbols." In *Principia*, the thesis is advanced as plausible but not proven.

The core of the present reconstruction is a weaker thesis which is arguably implicit in *Principia*, even if it is not as central to the work as it is to my reconstruction of it. The thesis may be formulated as the *epistemic dependency thesis*:

Except for finite classes, our knowledge of a class cannot consist in knowledge of its members but must appeal to a propositional function which the members of the class all satisfy.

The epistemic dependency thesis depends on the semantic dependency thesis for its generality, since if the latter thesis failed to hold for some infinite class, the class would not be epistemically accessible. Although the converse dependence does not hold, the semantic dependency thesis is unmotivated outside the context of a logical theory of classes. It seems plausible to suppose that any *logical* theory of classes is committed to the epistemic dependency thesis.

By the *no-classes theory of classes* I understand the philosophical proposal that we should infer from the truth of the semantic dependency thesis that classes are superfluous. The semantic dependency thesis is the basic premise of the no-classes theory, and in *Principia* the superfluousness of classes is derived from the reducibility of truths concerning classes to truths involving the use of class abstracts and a principle of ontological economy:

we shall assume a proposition about a class always to be reduced to a statement about a function which defines the class, *i.e.* about a function which is satisfied by the members of the class and no other arguments. Thus a class is an object derived from a function and presupposing the function, just as, for example, $(u).φu$ presupposes the function $φû$.[6]

In the case of descriptions it was possible to *prove* that they are incomplete symbols. In the case of classes, we do not know of any equally definite proof. It is

[6] *PM*, pp. 62–3; unqualified references to *Principia Mathematica* such as this one are to the first volume.

not necessary for our purposes to assert dogmatically that there are no such things as classes. It is only necessary for us to show that the incomplete symbols which we introduce as representative of classes yield all the propositions for the sake of which classes might be thought essential. When this has been shown, the mere principle of economy of primitive ideas leads to the non-introduction of classes except as incomplete symbols. (*PM*, p. 72)

Indeed, the no-classes theory is so closely linked to its justification in terms of the semantic dependency thesis that the two are often identified. The basic contention of the present reconstruction is that we should refrain from drawing an ontological lesson from the semantic dependency thesis but, in accordance with the epistemic dependency thesis, take our knowledge of a class to be facilitated by our knowledge of an appropriate propositional function.

There is a third dependency thesis associated with the logical theory of classes that I have not mentioned:

(*) A class *constitutes one object* or *forms a unity* because its elements all satisfy a common propositional function.

This thesis does not play a major role in my analysis. To begin with, the logical theory of classes seeks to *explain* the unity of a class by appeal to the class's association with a propositional function. But if we assume the existence of classes, it is unclear how explanatory (*) is. For then, given a class α together with the relation ϵ of class membership, (*) is satisfiable by the function $\hat{u} \, \epsilon \, \alpha$. By contrast with the *logical* notion of *class*, the *mathematical* notion of *set* holds not only that sets are extensional, but that they are *constituted* by their members. Classes also satisfy a principle of extensionality (*PM*, *20.15), but they are not constituted by their elements. A set may comprise some sort of unity, but on the mathematical conception, this is not explained by its association with a propositional function, but has the status of a primitive fact about sets and the process of collecting. The difficulty raised by appeal to a function like $\hat{u} \, \epsilon \, \alpha$ is that it does not take us any further than the mathematical concept of set toward explaining the unity of α. At most (*) shows that the logical conception of class can be justified relative to the mathematical notion of set, an observation which I take to be the correct lesson of *Principia*'s discussion of the assumption that there are classes.[7]

---

[7] Cf. *PM*, p. 58:

   It should be observed, in the first place, that if we assume the existence of classes, the axiom of reducibility can be proved. For in that case, given any function $\varphi\hat{u}$ of whatever order, there is a class consisting of just those objects which satisfy $\varphi\hat{u}$. Hence 'φu' is equivalent to 'u belongs to α.' But 'u belongs to α' is a statement containing no apparent variable, and is therefore a predicative function of u.

Suppose, however, that the italicized expressions of (*) are replaced with *constitutes one object of thought* and *forms a unity capable of entering into a judgment*, respectively; this yields:

(**) A class *constitutes one object of thought* or *forms a unity capable of entering into a judgment* because its elements all satisfy a common propositional function.

(**) follows from the epistemic dependency thesis. It also may appear to be trivially satisfiable since, for any given class, we can always express such a function with a name for the class and the symbol for the membership relation. However there is a reasonable basis for rejecting the suggestion that the thesis is so easily satisfied in its epistemic form. To satisfy (**) the class must be "given" to us in order to be provided with a name from which an expression for a function for it can be constructed. But if the class is sufficiently complex, it may not be possible that it should be given to us as the argument assumes: (**) may be satisfiable – there may be a function by which the class can be known – but that there is, is not established by so simple an extension of the argument that sufficed for (*).

10.4 THE CONCEPT OF A PROPOSITIONAL FUNCTION

In light of the foregoing, what principles should constrain the concept of a propositional function and the notion of class that it supports? There is an aspect of the iterative concept of set that bears on our question. Although it may seem strange to apply an idea from the mathematical concept of set to the rational reconstruction of a logical theory of classes, it will soon become apparent that such an application is not only legitimate, but is actually mandated by the consistent development of the concept of a propositional function and the theory to which it belongs. The aspect of the iterative concept of set that we require is the independence it ascribes to the elements of a set: according to the iterative concept, a set's elements exist independently of their membership in the set. A parallel supposition constrains the concept of a propositional function:

Argument Independence. The objects on which a propositional function acts exist independently of their connection with the function.

This observation is a kind of consistency proof of reducibility relative to a theory of classes. I say "a kind of consistency proof," since Whitehead and Russell assume that the language of *Principia* is indefinitely extendible, so that for every class there is a corresponding predicative function and predicative function-expression. Whitehead and Russell use the observation to argue that reducibility is a "smaller" assumption than the existence of classes.

There is some textual basis for *Argument Independence*, at least for the case of functions of individuals. Thus, in the 1910 introduction an argument is said to be a "constituent" of the proposition which is the value of the function for it as argument. The explanation of constituency which immediately follows has the clear implication that constituents exist independently of the function and are properly objects of quantification, or what Russell calls "complete symbols":

> objects which are neither propositions nor functions ... we shall call *individuals*. Such objects will be constituents of propositions or functions, and will be *genuine* constituents, in the sense that they do not disappear under analysis, as for example classes do, or phrases of the form 'the so and so.' (*PM*, p. 51)

For sets, two ideas are critical. The first idea, typically associated with the iterative conception, is the elements' independence of any set to which they belong. The second idea is the *dependence* of a set on its elements – the *extensionality* of sets; this is essential to *any* concept of set, but, as we have noted, on the iterative conception extensionality is forced by the requirement that a set is *constituted by* its elements. By contrast, propositional functions are not constituted by the things of which they are true; they are therefore not required to satisfy an extensional criterion of identity.[8] Propositional functions, unlike sets, fulfill a *representational* role. This comes about as follows.

A central problem which *Principles of Mathematics* (Russell 1903) sought to address is to explain how it is possible to have knowledge of objects of infinite complexity when our intelligence is capable of grasping only objects of finite complexity. Its answer depends on the theory of denoting concepts:

> With regard to infinite classes, say the class of numbers, it is to be observed that the concept *all numbers*, though not itself infinitely complex, yet denotes an infinitely complex object. This is the inmost secret of our power to deal with infinity. An infinitely complex concept, though there may be such, certainly cannot be manipulated by the human intelligence, but infinite collections, owing to the notion of denoting, can be manipulated without introducing any concepts of infinite complexity. (*Principles*, sect. 72)

Principia's account "of our power to deal with infinity" differs from that of *Principles* because of *Principia*'s almost total rejection of the theory of

---

[8] A point of difference between propositional functions and Fregean concepts is that the latter satisfy a principle of extensionality, a condition Frege took to be weaker than the claim that concepts are constituted by the objects which fall under them. For Frege even classes are not constituted by their elements, as is clear from his remark to Peano that "one must not view a class as constituted by its objects for then in removing the objects one would remove the class" (from an undated but post-1891 letter from Frege to Peano; collected in McGuinness 1980, p. 109).

denoting on which *Principles* is based. In *Principia*, the relevant entities of finite complexity are propositional functions. By grasping them we are able to have knowledge of infinitely complex objects. But the feature of propositional functions on which their ability to accomplish this task depends is the presence in them of variables; it is this feature that distinguishes functions from sets and classes and it is the presence of variables in functions that facilitates their finitary representation of objects of infinite complexity.[9]

Variables are acceptable in a way in which the denoting concepts of the older theories of denoting are not. Denoting concepts – such as *all men*, *some numbers*, and *the even prime* – differ from one another both formally and materially. However, variables are distinguished from one another only formally. The association of a variable with its range shares with the older notion of denoting the idea that a variable is *symbolic*, or, as I prefer to say, *representational*. But other aspects of denoting that are appropriate to denoting concepts are foreign to the relation a variable bears to its range. For example, in the older theory, different denoting concepts are associated with different pluralities of individuals, and within a plurality, with different combinations of its members. Thus *all numbers* denotes a different non-relational combination of the plurality of numbers than does *some numbers*, while *the even prime* denotes an individual rather than any kind of combination of members of a plurality. Nothing like this is true of variables; their distinctive feature is that they are capable of standing indifferently for anything falling within their range. *Principia* continues to use the language of the theory of denoting, saying of $\varphi\hat{u}$ that it ambiguously denotes φa, φb, φc, etc. by virtue of the presence of the variable u. The relation of *ambiguously denoting* or *ranging over* that holds between a variable and its range replaces the earlier relation of denoting which a denoting concept bears to its denotation.

Since it cannot be taken for granted that variables are unrestricted, it is necessary to make allowance for the possibility that a function's component variables have limited ranges. It follows that the domain of possible arguments to a function may also be restricted. Although propositional functions are not constituted by the things of which they are true, the presence of variables in propositional functions means that they satisfy a dependence condition that is weaker than – but evidently related to – the principle of

---

[9] In this connection compare Russell's remark to Moore that "[in 'On denoting'] I only profess to reduce the problem of denoting to the problem of the variable" (letter to Moore of October 25, 1905, quoted in the introduction to Russell 1994, p. xxxv). I agree with Russell, but I do not treat variables as denoting concepts; Russell's cautious formulation suggests that at the time of writing he may not have done so either.

extensionality. I call this condition *Function Dependence* to distinguish it from the complete dependence of a set on its elements, and I formulate it as follows:

Function Dependence. A propositional function presupposes the totality of its possible arguments and any totality over which its constituent quantified variables range.[10]

The imposition of *Function Dependence* is justified merely by the requirement that there should be some provision for the possibility that functions are in some way typed; the condition does not, by itself, impose such a restriction.

Let me briefly summarize the conclusions we have reached: the logical notion of class is based on that of a propositional function. And the relevant concept of propositional function is constrained by *Argument Independence* and *Function Dependence*. These constraints parallel those governing the mathematical concept of set: *Argument Independence* is the analogue of the thesis that sets are *based* or *founded on* their elements and *Function Dependence* is the analogue of the thesis that sets are *constituted* by their elements.

As for the vicious circle principles that constituted Russell's mature diagnosis of the paradoxes and led him to ramification, we wish to know whether they are analytic of the notions of *propositional function* and *class* as we have just proposed these notions should be understood.

Following on the discussion of Gödel (1944), it is useful to distinguish a general vicious circle principle (VCP) for "totalities," and two that apply more specifically to functions:

> *General VCP.* No totality can contain members involving or presupposing the totality; hence, no totality of all arguments to a function can contain members involving or presupposing the totality of its arguments.
>
> *Strong VCP.* No function can belong to a totality which it involves or presupposes.
>
> *Weak VCP.* No function can have arguments involving or presupposing the function.

All three principles are distinct from the prohibition against impredicative definition for which there are three parallel formulations:

---

[10] A representative formulation is: "A function ... presupposes as part of its meaning the totality of its values, or, what comes to the same thing, the totality of its possible arguments" (*PM*, p. 54). And "the values of a function are presupposed by the function, not vice versa" (*PM*, p. 39).

General Impredicativity. No totality can contain members definable only in terms of the totality; hence, no totality of all arguments to a function can contain members definable only in terms of the totality of its arguments.

Strong Impredicativity. No function can belong to a totality which is involved in its definition.

Weak Impredicativity. No function can have arguments definable only in terms of the function.

My immediate concern is with the first group of vicious circle principles; I will take up the "impredicativity principles" later.

Weak VCP follows immediately from *Function Dependence*, but it is also a simple consequence of just *Argument Independence*: if our concept of a propositional function is such that its arguments do not depend on the function itself, then functions cannot be involved in or presupposed by their arguments without contravening *Argument Independence*. By contrast with *Weak VCP*, *Strong VCP* requires the full strength of *Function Dependence*. Turning to *General VCP*, we can show that it holds for those totalities that are classes, and, a fortiori, for those totalities of arguments to a function that are classes. Indeed, for such totalities, *General VCP* is a consequence of the *semantic dependency thesis* and *Argument Independence*. For, by the semantic dependency thesis any truth concerning classes is transformable into one in which reference to the class is replaced by reference to a propositional function that determines it. In particular, if there were a truth concerning a class that contained a member involving or presupposing the class, there would be a corresponding truth concerning a propositional function one of whose arguments involved or presupposed the function. But this is impossible by *Argument Independence*.

Turning to the impredicativity principles, Gödel (1944) argued that from a realist perspective on the totalities comprised by *sets*, the imposition of *General Impredicativity* is not justified. From such a perspective, a definition is merely an expression that uniquely specifies the defined entity. There is therefore nothing problematic in the notion that our only means of identifying a set might involve quantification over a totality to which it itself belongs. Gödel argued that for an impredicative definition to be problematic, one must think of the definition as not merely uniquely picking out the entity defined, but as somehow essential to the entity's existence. This would be plausible if, for example, a definition were thought of as an explanation of how to effect a construction of what is defined. For it could then be argued that an entity cannot be assumed in its own construction. But if we conceive of sets as existing independently of our constructions, there is no reason to accept *General Impredicativity*.

In response to this argument of Gödel's, Goldfarb (1989) observed that the defense of impredicative definitions of sets depends on an assumption much less tendentious than realism. A set is an extensional entity whose members are essential to it; but no defining formula is essential to the identity of the set. Sets are not preserved under change of elements, but it is a matter of indifference to the identity of a set that it is specified by a defining formula that contains a quantified variable whose range includes the set itself. The prohibition against impredicative definitions of sets therefore lacks any intuitive basis even if one is not a realist regarding their existence.

Goldfarb's observation is certainly correct. The point might be put by saying that the acceptability of *General Impredicativity* turns on a commitment regarding the *essence* of sets, rather than one regarding their *existence*. The nature of the basis for the extensionality that sets enjoy suffices to show why one might plausibly reject *General Impredicativity* for them. But Gödel's chief contention was that only a form of *anti*-realism could sustain the imposition of *General Impredicativity* on *propositional functions*. As Goldfarb recognizes (1989, p. 31), his observation does not by itself refute this contention of Gödel's, but shows only that since propositional functions are not extensional entities, they *may* be constrained by *General Impredicativity*, not that they *must* be. Goldfarb's proposal for addressing this lacuna is to argue that our access to a propositional function is altogether different from the access we have to a class specified by the function:

> To specify a class is to give a propositional function that is true of all and only the members of the class. The specification must be understood on its own; given such an understanding, it is a further question whether or not a given class is the one specified ... [But] the comprehension axioms for propositional functions that are implicit in [*Principia*] involve not so much the specification of these entities as the presentation of them. I shall therefore use 'presentation' rather than 'specification' in this connection. (Goldfarb 1989, pp. 31–2)

Goldfarb's idea is that with the specification of a class there is a certain "space" between the specification and the class specified. But to "specify" a propositional function is to be "immediately presented with" the entity meant. The suggestion is an interesting one, but it is difficult to see how to make entirely explicit the notion of immediate presentation on which it depends. It may therefore be worthwhile to see if there is an alternative justification which avoids the notion of immediate presentation.

The strategy we followed for *General, Strong*, and *Weak VCP* was to argue that they are forced by the concept of a propositional function. This strategy appears to be correct for the impredicativity principles controlling functions, since a definition which violates *Strong* or *Weak Impredicativity* can be excluded on the ground that the function it purports to express would violate *Strong* and *Weak VCP*. But *General Impredicativity* has a different justification, one that depends on the centrality of propositional functions to the explanation of our knowledge of classes. According to the logical theory, an expression that defines a class in terms of a totality to which the defining function itself belongs fails to define it by a propositional function, since a function given by such a definition is incapable of supporting an explanation of how the class is known.

Consider, again, the celebrated Frege–Dedekind definition of the natural numbers as the class determined by the function that holds of those individuals which satisfy all inductive functions of zero. This definition fails to express a propositional function because it allows a function to fall within the range of one of its quantified variables. Consequently, what the definition expresses violates both *Weak* and *Strong VCP*. The rationale for *General Impredicativity* is therefore that the definitions it excludes fail to express propositional functions, and hence cannot support the *epistemic dependency thesis*. But the *justification* for imposing *General Impredicativity* can only be achieved by demonstrating the adequacy of the hierarchy of classes that *are* capable of being known by propositional functions which accord with *Strong* and *Weak VCP*. This brings us to the Axiom of Reducibility.

10.5 REDUCIBILITY

On the present reconstruction, the Axioms of Reducibility ensure the epistemic accessibility of certain classes. The axioms are formulated with typical ambiguity and occur at every type.[11] They postulate that the quantified variables of a propositional function range over predicative extensional equivalents of functions of every order within a type. According to the ramified theory, a function is prohibited from falling within the range of one of its quantified variables, but the theory imposes no such restriction on functions *coextensive* with it. The point of the 1910 theory is epistemic and the Axioms of Reducibility secure our knowledge of particular classes in terms of *nonpredicative* functions by the postulation of the requisite predicative functions.

---

[11] The simplest example of an Axiom of Reducibility involves the postulation of predicative equivalents of propositional functions of a single argument; formally, $\exists\chi\forall u(\varphi u \equiv \chi!u)$, where the exclamation mark indicates that $\chi\hat{u}$ is a predicative function.

This epistemological role of reducibility has been missed, even by some of *Principia*'s most astute commentators. Thus it has been noted of Gödel that his

remarks about the axiom of reducibility show [a] lack of sensitivity to the essentially intensional character of Russell's logic; the fact that every propositional function is *coextensive* with one of lowest order does not imply "the existence in the data of the kind of objects to be constructed" (1944, p. 141), if the objects in question are . . . propositional functions rather than classes . . . The ramified hierarchy with reducibility would fit with a conception according to which classes are admitted as "real objects," but the conception of propositional function is . . . constructivistic. (Parsons 1990, pp. 112–13)[12]

While the postulation of predicative equivalents of functions of all orders significantly affects the totality of classes that are determined by any non-predicative function, the inclusion of predicative equivalents in the range of such a function's variables does not require any modification of *Principia*'s account of our knowledge of the propositional function itself. This is because its account of our knowledge of functions and classes depends on an understanding of generality according to which the use of a variable does not require knowledge of its values. There is, in particular, no presumption that in order to grasp a propositional function, it is necessary to know any function or individual in the range of a variable to the function:

[A] function can be apprehended without its being necessary to apprehend its values severally or individually. If this were not the case, no function could be apprehended at all, since the number of values (true and false) of a function is necessarily infinite and there are necessarily possible arguments with which we are unacquainted. (*PM*, pp. 39–40)

In *The Problems of Philosophy*, Russell devotes a great deal of space to the defense of this account of general knowledge, seeking to establish that much of what appears to require knowledge of particular instances does not in fact do so. The problem becomes especially pressing in connection with a priori knowledge, since it is supposed to be independent of knowledge of particular facts of experience and of the existence of particular individuals. Russell argues that we can understand a proposition purporting to express a priori knowledge without knowing the constituents of any instance of the proposition, and we can do so compatibly with an empiricist commitment to the primacy of acquaintance and the empiricist thesis that all knowledge of

---

[12] The reconstruction of ramification plus reducibility proposed here is one possible development of Parsons' suggestion, but I do not pretend to know if it conforms to what he had in mind or whether he would agree with it.

particular existence is a posteriori. There are even cases of a posteriori known general truths that do not depend on knowledge of their instances. Russell's example is the following proposition:

(§) There are numbers greater than 1000 which no one has ever thought of.

By the nature of the case we cannot know an instance of (§), since that would require knowing the number which is a constituent of the instance, thereby precluding the instance from being a witness to the truth of (§).

This suggests a novel explanation of the significance of Russell's theory of descriptions for his subsequent logical discoveries. The interest of that theory is usually restricted to its application to the semantic issues posed by vacuous descriptions. From such a perspective, analyzing classes away after the fashion of the contextual analysis of vacuous singular terms is naturally viewed as the principal lesson the theory has to offer for *Principia*'s account of functions and classes. But by focusing on the elimination of the bearers of vacuous descriptions – as the no-class theory does – it is easy to lose sight of the theory of descriptions' equally important epistemic applications – applications that arise in cases where there is no question of failure of reference.

Principia's theory of knowledge of classes parallels Russell's general theory of knowledge of things; *Principia*'s theory would also collapse if our understanding of generality demanded knowledge of instances. Classes stand to the functions which determine them as individual things stand to the descriptive functions they satisfy. Just as there are many descriptions of the same individual, so also the same class is capable of being the extension of many functions. Knowledge of a class by means of a predicative function corresponds in the theory of knowledge of things to being acquainted with an individual and in possession of a logically proper name for it. Knowledge of a class by means of a nonpredicative function is the analogue of knowledge of an individual by means of a description.

The logical complexity of a nonpredicative function follows its canonical linguistic expression in the language of *Principia*; knowledge of the function proceeds from knowledge of its predicative components and the logical form of their combination. In the case of knowledge of things, the transparency of the components of the description by which an individual is specified is *epistemic*. The components of the description (or rather, the components of the propositional function which replaces it) are known by acquaintance. But although the components are known by acquaintance, the description to which they belong may be satisfied by something that transcends our acquaintance.

By the epistemic dependency thesis, our access to classes is through propositional functions. A class for which we lack a predicative function is accessible only if it is determined by a logical construction built from known predicative functions. In the case of classes, the relevant transparency of the components of the functions by which they are known is their *logical* transparency, and it is represented by the fact that the basic functional constituents are predicative functions. In analogy with the theory of knowledge of things which transcend our acquaintance, although the basic component propositional functions are predicative, the class determined by the logical construction which they comprise can be one that is not known by means of a predicative function.

It is difficult to see the point of a theory consisting of ramified types plus reducibility if it is detached from its epistemological motivation. Nor is it easy to see in what sense the reducibility axioms represent an advance over the mere postulation of classes. But they represent such an advance in at least two respects: they preserve a logical theory of our knowledge of classes, according to which classes are known as the extensions of propositional functions; and the reducibility axioms also preserve an attractive representation of our knowledge of particular classes in terms of nonpredicative functions. In connection with this second advance, the paradigm case of such a representation is the impredicative definition of the class of natural numbers of Frege and Dedekind. Although it is not required that we should know the values of a variable occupying the argument-place of a function in order to know the class the function determines, it *is* required that the range of its values *be* sufficiently extensive in order that the function should determine the intended class.

The ramified analogue of the Frege–Dedekind definition captures the class of natural numbers only through the intervention of the Axiom of Reducibility. It is for this reason that the correctness of *Principia*'s reformulation of the definition has been sometimes thought to capture the class of natural numbers only "accidentally," by the coincidence that the range of the variable which occurs in it is *as a matter of fact* sufficiently extensive. Such contingency at the foundational level is characteristic of *Principia*'s reconstruction of arithmetic, and is perhaps most conspicuous in its formulation of its Axiom of Infinity.[13] As we have just seen, it also arises in connection with reducibility. But contingency is not ad hocness, and whatever *Principia*'s defects, it has the virtue of being explicit about the postulates that honest logical toil seems to require.

---

[13] See Section 1.3 for a discussion of the bearing of *Principia*'s Axiom of Infinity on its account of arithmetical knowledge.

APPENDIX: RUSSELL'S PROPOSITIONAL PARADOX[14]

In the final appendix to *The Principles of Mathematics*, Russell derives a paradox that has come to be known as "the propositional paradox." *Principles* contains the only mention of this paradox in Russell's published writings, and his discussion of it contains the earliest anticipation of his idea that a unitary solution to all the paradoxes might require something more than the simple theory of types which he had sketched in an earlier appendix. The propositional paradox is not "mathematical" but neither is it clearly semantic. Like Russell's paradox of the class of all classes that are not members of themselves, the propositional paradox illustrates how "[f]rom Cantor's proposition that any class contains more subclasses than objects we can elicit constantly new contradictions."[14] While one can extract from the derivation of the paradox a proof that there are more properties of propositions than there are propositions,[15] its main interest, like the interest of the Russell–Zermelo paradox of the class of all classes that are not members of themselves, consists in the fact that it challenges a number of basic assumptions, in this case, assumptions about the nature of propositions and propositional identity. By an extension we will explore presently, these assumptions bear on the theory of Fregean thoughts and Dedekind's account of our acquisition of the concept of number.

The formulation of the paradox I will follow[16] focuses on propositions of the form

$$\forall p(\varphi(p) \rightarrow p).$$

In a propositional theory like Russell's, such a proposition is understood as saying that every proposition with property φ is true. Notice that in order for '$\varphi(p)$' to express a propositional function, and for '$\varphi(p) \rightarrow p$' to express a proper truth function of $\varphi(p)$ and p, 'p' must have what I will call a *designative occurrence* in '$\varphi(p)$': i.e. substituends for 'p' must designate propositions. By contrast, the occurrence of 'p' in the consequent of '$\varphi(p) \rightarrow p$' must not be designative, but *expressive*, i.e. a substituend for 'p' must express a proposition. There is a systematic ambiguity between designative

---

[14] See Russell's letter to Frege of September 29, 1902 in McGuinness (1980, p. 147).

[15] See Linsky (1999) for a discussion of the paradox along these lines.

[16] The paradox is discussed at length in Russell's correspondence with Frege. See especially the series of exchanges which begins with Russell's letter of September 29, 1902 and concludes with his letter of May 24, 1903. I first became convinced of the philosophical interest of the paradox from reading about it in Potter (2000). The exposition which follows draws on the discussion of the paradox in Myhill (1958), Church (1984), Potter (2000), and Klement (2001 and 2002).

and expressive occurrences of 'p' in the argument to the paradox, but there is no confusion of designative and expressive occurrences. As remarked by Church,[17] the paradox is derivable for a wide choice of propositions. In the extension to Frege's theory of thoughts given below, we will derive the paradox using a proposition whose universal quantifier binds only designative occurrences of 'p.'

Although, as we have noted, '$\forall p(\varphi(p) \to p)$' is naturally taken to express the proposition that every proposition with property φ is true, the notion of a truth *predicate* does not enter into the derivation of the paradox. Nor does the paradox require for its formulation the availability of a semantic relation of designation, a point first made explicit by Church who derived the paradox within a theory consisting of the simple theory of types augmented by the axioms of propositional identity that are implicit in *Principles*. Thus, in terms of Ramsey's well-known division of the paradoxes into those belonging to logic and mathematics "in their role as symbolic systems" versus those which depend on the bearing of logic on "the analysis of thought," the propositional paradox falls squarely into the second category; yet, for the reasons just given, it is not straightforwardly *semantic*.

The paradox arises as follows. Let φ be a property of propositions; in Russell's terminology, φ is a propositional function which takes propositions as arguments and yields propositions as values. Consider the function f from properties of propositions to propositions defined by

$$f\varphi =_{\mathrm{Df}} \forall p(\varphi(p) \to p).$$

The notation is justified by Russell's theory of propositions: Since φ is the only property of propositions which is a constituent of $f\varphi$, f is a many–one relation, and on the assumption that for every φ there is such a proposition $f\varphi$, f is a function. From the fact that φ is a constituent of the proposition $f\varphi$, it follows by Russell's conception of propositions and propositional identity (see Church, p. 520) that

$$(f\varphi = f\psi) \to (\varphi = \psi),$$

a consequence we will call *Injectivity*. Now let ψ be the property of propositions given by

$$\psi(p) =_{\mathrm{Df}} \exists \varphi(p = f\varphi \wedge \neg\varphi(p))$$

---

[17] Church (1984, p. 520). All references to Church are to this paper.

and suppose $\neg\psi(f\psi)$:

$$\neg\exists\varphi(f\psi = f\varphi \wedge \neg\varphi(f\psi));$$

i.e.

$$\forall\varphi(f\psi = f\varphi \to \varphi(f\psi)),$$

and therefore, $\psi(f\psi)$.

Next suppose $\psi(f\psi)$:

$$\exists\varphi(f\psi = f\varphi \wedge \neg\varphi(f\psi)).$$

Then by *Injectivity*,

$$\exists\varphi(\varphi = \psi \wedge \neg\varphi(f\psi)),$$

whence $\neg\psi(f\psi)$.

The paradox requires the theory of Russellian propositions developed in *Principles* to motivate *Injectivity*. In the context of Russell's conception of propositions, *Injectivity* is forced by the assumption that the propositional function φ is a constituent of $f\varphi$. The assumption that for every φ there is a proposition $f\varphi$ – namely, $\forall p(\varphi(p) \to p)$ – is so intuitively plausible that it would seem to be a necessary component of any theory of propositions. And on the assumption that for every φ there is a proposition of the form $\forall p(\varphi(p) \to p)$, the derivation requires only that the relation which associates $\forall p(\varphi(p) \to p)$ with φ – the inverse of the relation determined by f – should be functional, a point to which we will return in the context of Frege's theory of thoughts.

The solution mandated by the ramified theory of types depends on observing the restrictions it imposes on the "orders" of the propositions and propositional functions to which the derivation of the paradox appeals. The technical idea behind the ramified theory's stratification of *propositions* into orders is that the order of a proposition p must exceed the order of any proposition that falls within the range of one of its quantified propositional variables, and therefore, by the restrictions of the theory, p cannot itself fall within the range of one of its quantified propositional variables. By an extension of terminology, the order of a propositional variable is the order of the propositions which belong to its range of values. According to the ramified theory, the proposition $q =_{\mathrm{Df}} \forall p(\psi(p) \to p)$ and the propositional function ψ, defined in the course of the derivation of the paradox, are such that if the order of the bound propositional variable in q is n, the orders of q

and ψ must be $n + 1$, so that q is not itself a possible argument for the function ψ. Since the contradiction was derived from the supposition that $ψ(q)$ or $¬ψ(q)$, the argument collapses once the restrictions imposed by ramification are taken into account. Notice that this solution preserves the idea that φ is a constituent of $fφ$, as well as the idea that there always is a proposition $fφ$ corresponding to φ. However, the generality of these assumptions is the restricted generality of the ramified theory: for every ψ of order $n + 1$, there is a proposition of order $n + 1$ which asserts the truth of every proposition of order n for which ψ holds.

As is well known, there are two hierarchies that emerge from Frege's philosophical reflections on logic and language: the hierarchy of functions (or concepts) and objects, and the hierarchy of senses. The first hierarchy has a simple type-theoretical structure and, like the simple theory of types, is naturally motivated by considerations of predicability: first-level functions are predicated of objects, second-level functions of first-level functions, third-level functions of second-level functions, etc. There is therefore no analogue in Frege's hierarchy of functions of Russell's paradox of the propositional function that is predicable of all and only those propositional functions which are not predicable of themselves. But as we have seen, the propositional paradox can be derived from the theory of Russellian propositions of *Principles* even when this theory is formulated within a simple type-theoretical formalism. It is therefore natural to ask whether Frege's theory of sense, and in particular, his theory of thoughts – which are just a subclass of senses – fares just as badly: to ask in other words whether Fregean thoughts, which are the proper correlates of Russellian propositions, admit the construction of an analogue of the propositional paradox.

Notice that in so far as senses are objects, they too fall under concepts. Thoughts, being the senses of complete expressions, are objects. Among thoughts, some are *universal*, i.e. are composed of the senses for the universal quantifier and a concept-expression. Let us call the concept to which such a concept-expression refers, the *principal concept of the universal thought*. Although Frege's hierarchy of *concepts* escapes Russell's paradox of the predicate which holds of all predicates that are not predicable of themselves, there appears to be nothing to prohibit the concept χ that holds of all universal *thoughts* that do not fall under their principal concept. Does the universal thought expressed by, for example, 'Everything is χ' fall under χ? If it does, it violates the defining characteristic of χ, and so does *not* fall under χ. And if it does not, it violates the universality of χ, and therefore, *does* fall under χ.

The informal argument we have just given goes back to Myhill (1958). More explicitly, let h be a relation which relates every concept φ to a Fregean thought: the universal thought $\forall p(\varphi(p))$ of which it is the principal concept. Thus, $\varphi h q$ if, and only if, $q = \forall p(\varphi(p))$. There may of course be distinct universal thoughts whose constituent senses determine φ as their principal concept. But at the level of the *senses* φ_s of concepts φ, the situation for Fregean thoughts exactly parallels Russellian propositions and propositional functions: senses being constituents of Fregean thoughts, the theory implies that the relation between universal thoughts and the senses of their principal concepts one–one maps the domain of such senses on to the class of universal thoughts.

The derivation of a paradox in Frege's theory requires that h satisfy the following two conditions: (i) every universal thought is in the range of the relation h, and (ii) whenever $\varphi h p$, $\psi h q$, and $p = q$, $\varphi = \psi$. It is clear that (i) is satisfied. As for (ii), if the universal thoughts p and q are the same, so also are their constituent senses φ_s and ψ_s, and since sense determines reference, their principal concepts must also be the same.

Now let $q = \forall p(\chi(p))$, where

$$\chi(p) =_{Df} \exists \varphi(\varphi h p \wedge \neg \varphi(p)),$$

then $\chi h q$. Now suppose that $\chi(q)$, i.e.

$$\exists \varphi(\varphi h q \wedge \neg \varphi(q)).$$

It then follows that

$$\exists \varphi(\chi h q \wedge \varphi h q \wedge \neg \varphi(q)),$$

so that using (ii),

$$\exists \varphi(\chi = \varphi \wedge \neg \varphi(q)),$$

whence, $\neg \chi(q)$.

Next suppose $\neg \chi(q)$:

$$\forall \varphi(\varphi h q \to \varphi(q)).$$

Thus, in particular, $\chi h q \to \chi(q)$. Since $q = \forall p(\chi(p))$, $\chi h q$ and therefore, $\chi(q)$.

The essential similarity between the concept–class and concept–thought paradoxes can be made intuitively compelling as follows: recall that an object belongs to a class if it falls under the concept whose extension the class is. Similarly, let us say that an object *belongs to a universal thought* if it falls under the principal concept of the thought. Then the concepts expressed by '*x* is the class of all classes that do not belong to themselves' and '*x* is the universal thought of all universal thoughts that do not belong to themselves' both lack objectual representatives. Frege shows[18] that, given any map *f* from concepts to objects, we can always find concepts φ and ψ such that $f\varphi = f\psi$ but $\varphi(f\varphi) \land \neg\psi(f\varphi)$. Thus, the fact that there cannot be an injection of concepts into objects is entailed by a consistent fragment of *Grundgesetze*, and this shows that the difficulty with Frege's theory of classes is an instance of a more fundamental difficulty, one that arises in his general theory of concepts and objects.

In the case of thoughts, Frege might argue that they, at least, are not *proper* objects – not arguments to concepts of first level. Thoughts are *like* objects in being "saturated" – in corresponding to "complete" expressions – but this is only one component of what characterizes something as an object. The most important component of being an object – that of being an argument to a concept of first level – may not be a feature thoughts enjoy; rather, they might form a different hierarchy from concepts and objects proper. But an examination of the argument leading to the paradox shows it to require only that thoughts fall under the concepts appropriate to them and is unaffected if thoughts and the concepts they fall under form a separate hierarchy from ordinary objects and concepts.[19]

Frege's celebrated proof of the Dedekind infinity of the numbers from their criterion of identity does not require any assumptions about the objectual character of thoughts, but rests on the thesis that *numbers* are objects.[20] Frege's argument is therefore not compromised by the extension of the propositional paradox to the concept–thought paradox. However, it is interesting to consider how Dedekind's logicism fares from the perspective of the propositional paradox. As Dummett (1991a, pp. 48–54) has observed, Dedekind's account of our acquisition of the concept of number requires the provision of a concrete example of a Dedekind-infinite system. Given the existence of such an example, Dedekind can argue that the concept of number is *abstracted* from it; and by appealing to his categoricity theorem (Dedekind 1888, Theorem 132), he can argue that the peculiarities

---

[18] See the appendix Frege added to *Grundgesetze* in response to Russell's letter of June 16, 1902 informing him of the contradiction.

[19] Cf. Dummett's discussion of Myhill's paradox in Dummett (2007, pp. 124–6).

[20] See Chapters 1 and 9, above; the discussion in Section 1.3 is particularly relevant to the remarks on Frege which follow.

of the example may be discounted, and the *generality* of the concept thus arrived at secured. But the provision of a concrete example is essential to Dedekind's strategy; its provision is the point of Theorem 66, whose proof runs as follows:

> The world of my thoughts, i.e., the totality S of all things that can be objects of my thought, is infinite. For if s denotes an element of S, then the thought s′, that s can be an object of my thought, is itself an element of S. If s′ is regarded as the image φ(s) of the element s, then the mapping φ on S determined thereby has the property that its image S′ is a part of S and indeed S′ is a proper part of S, because there are elements in S (e.g., my own ego), which are different from every such thought s′ and are therefore not contained in S′. Finally it is clear that if a, b are different elements of S, then their images a′, b′ are also different, so that the mapping φ is distinct (similar). Consequently, S is infinite, q.e.d.

As Frege noted, Dedekind's realm of thoughts is an objective one, one that exists independently of us, and one to which our access can be taken for granted since thoughts are transparent to our reason:

> Dedekind, in theor. 66 of his book *Was sind und was sollen die Zahlen?*, uses ['thought'] as I do. For he is attempting there to prove that the totality of things that can be objects of his thinking is infinite . . . It is essential to the proof that the sentence 's can be an object of Dedekind's thinking' always expresses a thought when the letter 's' designates such an object. Now if, as Dedekind wishes to prove, there are infinitely many such objects s, there must also be infinitely many such thoughts [that s can be an object of his thinking]. Now presumably we shall not hurt Dedekind's feelings if we assume that he has not thought infinitely many thoughts. Equally he should not assume that others have already thought infinitely many thoughts which could be the objects of his thinking; for this would be to assume what was to be proved. Now if infinitely many thoughts have not yet been thought . . . it cannot be essential to a thought that it should be thought. And this is precisely what I am maintaining . . . So the validity of Dedekind's proofs rests on the assumption that thoughts obtain independently of our thinking.[21]

---

[21] Frege (1897, p. 136). In this connection, compare Dummett (2007, p. 126):

> What disqualifies thoughts from being full-fledged objects is that they are not self-subsistent: a thought could not exist unless at least one human being or other rational creature grasped it . . . Thoughts . . . do not form, as Frege may well have supposed, a determinate domain containing every thought that ever will or ever could be grasped or expressed, some waiting in vain because they never will be expressed or grasped. The domain of thoughts is indeterminate and constantly expanding. Quantification over it, therefore, cannot be explained in the classical manner as the outcome of scrutinising each element of the domain to determine whether it satisfies the predicate or not. It can be explained only in the intuitionist fashion, under which an existential statement is justified only by the production of an instance, and a universal one by a demonstration of its necessitation from the very notion of a thought.

For both Frege and Dedekind, thoughts are not only objects – possible arguments to first-level functions – they are *self-subsistent* objects. But for Dedekind the realm of thoughts yields an exemplar of an infinite system that is essential to his theory of our knowledge of number: the relation, x *is an object of my thought y*, is a one–one function which maps a subsystem of objects of my thought on to a proper part of itself, and it is from this exemplar that we *abstract* the concept of an infinite system, and thus, of number. But on the assumption that Dedekind's tacit theory of thoughts is essentially Frege's, the concept–thought paradox is an obstacle to the implementation of Dedekind's account of our acquisition of the concept of number in a way in which it is not for Frege's. For if the proof of Theorem 66 rests on an incoherent theory of thoughts, we lack an exemplar of a Dedekind-infinite system from which our concept of number can be abstracted. Dedekind cannot therefore claim to have provided an account of how we acquire the concept of number, still less how we acquire it on the basis of an abstraction from an infinite totality comprised of our thoughts. But whatever difficulties the concept–thought paradox holds for the theory of Fregean thoughts, Frege's account of how we come to the concept of number does not rest on the provision of an exemplar from which it is supposed to be abstracted. For Frege, we acquire the concept with our recognition that its application is constrained by the criterion of identity. That Frege–Dedekind thoughts form a demonstrably inconsistent totality is, therefore, a difficulty for Dedekind's explanation of our acquisition of the concept of number in a way that it is not for Frege's.

ACKNOWLEDGMENTS

I am grateful to the Social Sciences and Humanities Research Council of Canada for support of my research. Thanks to John Bell, Peter Clark, Michael Dummett, Solomon Feferman, Michael Friedman, Kevin Klement, Peter Koellner, and Erich Reck for stimulating discussions on the topics dealt with here.

CHAPTER 11

Ramsey's extensional propositional functions

11.1 INTRODUCTION

A paper I wrote with John Bell some years ago[1] was occasioned by an article of Hintikka and Sandu[2] that questioned whether Frege's understanding of second-order logic corresponded, in Frege's framework of functions and concepts, to what we would now regard as the *standard* interpretation of second-order logic, the interpretation that takes the domain of the function variables to be the full power set of the domain of individuals. Hintikka and Sandu maintained that it did not on the basis of a number of arguments, all of which they took to show that Frege favored some variety of nonstandard interpretation for which the domain of the function variables is something less than the characteristic functions of all subsets of the domain over which the individual variables range.

Although Hintikka and Sandu based their contention on many different considerations, the one to which they assigned the greatest weight was that Frege lacked the concept of an arbitrary correspondence from natural numbers to natural numbers (or from real numbers to real numbers).[3] They contrasted their view of Frege on the concept of a function with that of Dummett who, in *Frege: Philosophy of Language*,[4] had maintained that Frege's notion of a function coincides with the concept of an arbitrary correspondence:

Based on my paper "On logicist conceptions of functions and classes," which appeared in D. DeVidi, M. Hallett, and P. J. Clark (eds.), *Vintage Enthusiasms: Essays in Honour of John L. Bell* (Dordrecht: Springer, 2011): 3–18.

[1] Demopoulos and Bell (1993).   [2] Hintikka and Sandu (1992).
[3] The textual considerations Hintikka and Sandu advanced in support of this interpretive claim were shown by Burgess (1993) to rest on simple misreadings of Frege's *venia docendi* – Frege (1874) – and in the case of Frege (1904) – the relatively late paper, "What is a function?" – on passages that are far from unequivocal.
[4] I often use the following acronyms and abbreviations: *FPL* for Dummett (1981), *FPM* for Dummett (1991a), *Grundlagen* for Frege (1884), *Grundgesetze* for Frege (1893/1903), *FoM* for Ramsey (1925b), *Principia* for Whitehead and Russell (1910), and *Tractatus* for Wittgenstein (1922).

And it is true enough, in a sense, that, once we know what objects there are, then we also know what functions there are, at least, so long as we are prepared, as Frege was, to admit all "arbitrary functions" defined over all objects. (*FPL*, p. 177, quoted by Hintikka and Sandu, p. 303)

However, in *Frege: Philosophy of Mathematics* Dummett changed his mind on this question, and suggested that Frege implicitly assumed the adequacy of a substitutional interpretation of his function variables:

Frege fails to pay due attention to the fact that the introduction of the [class] abstraction operator brings with it, not only new singular terms, but an extension of the domain ... [The operator] may be seen as making an inconsistent demand on the size of the domain D, namely that, where D comprises n objects, we should have $n^n \leq n$, which holds only when $n = 1$, whereas we must have $n \geq 2$, since the two truth values are distinct: for there must be n^n extensionally nonequivalent functions of one argument and hence n^n distinct value ranges. But this assumes that the function-variables range over the entire classical totality of functions from D into D, and there is meagre evidence for attributing such a conception to Frege. His formulations make it more likely that he thought of his function-variables as ranging over only those functions that could be referred to by functional expressions of his symbolism (and thus over a denumerable totality of functions), and of the domain D of objects as comprising value-ranges only of such functions. (*FPM*, pp. 219–20)

Dummett[5] put forward a further consideration that can be seen as lending support to Hintikka and Sandu's view that Frege favored some sort of nonstandard interpretation. In the course of his discussion of the attempted consistency proof of sect. 31 of *Grundgesetze*, Dummett suggested that Frege erroneously supposed that the consistency of certain restricted interpretations of the function variables extends to the system of *Grundgesetze*'s theory of functions and classes in the context of *full* second-order logic. This unjustified, and unjustifiable, assumption is what Dummett means by "Frege's amazing insouciance regarding the second-order quantifier" (*FPM*, p. 218). Subsequent developments have shown that there is a systematic consideration in favor of Dummett's claim and, a fortiori, in favor of Hintikka and Sandu's suggestion that Frege assumed a nonstandard interpretation: some restricted interpretations yield consistent fragments of *Grundgesetze*.[6] Hence it might be that Frege was blind to the inconsistency of his system because he considered the question of consistency only from the perspective of such a restricted interpretation and failed to ask whether what holds of it, holds in general.

---

[5] In *FPM*, Chapter 17, "How the serpent entered Eden."
[6] See, for example, the system **PV** discussed in Burgess (2005, sect. 2.1).

But although Dummett shares Hintikka and Sandu's conclusion that Frege tended toward a nonstandard interpretation, his analysis does not support Hintikka and Sandu's evaluation of Frege's foundational contributions. If we follow Dummett, Frege missed the fact that the consistency of *Grundgesetze*, relative to a nonstandard interpretation, does not necessarily extend to its consistency when the logic is given a standard interpretation. This is certainly an oversight, but it is not the oversight that is appealed to in those of Hintikka and Sandu's criticisms of Frege that so offended some of their critics, as for example, whether, without having isolated the notion of a standard interpretation, Frege could have even conceptualized results like Dedekind's categoricity theorem.

Apropos of Frege and categoricity, in their response to Hintikka and Sandu, Heck and Stanley (1993) observed, what Heck (1995) later elaborated in detail, that Frege proved an analogue of Dedekind's theorem using his own axiomatization of arithmetic (one that is only a slight variant of the Peano–Dedekind axiomatization). And in our response, Bell and I argued that the relevance of the dependence of this and other similar foundational results on the standard interpretation is not entirely straightforward, since the actual arguments which support them have the same character, whether one is working in second-order logic or in a suitably rich first-order theory such as Zermelo–Fraenkel set theory. Hintikka and Sandu's claim that Frege could not even have formulated (let alone appreciated) these results because of their dependence on the standard interpretation is therefore incorrect both historically and methodologically. It is incorrect historically because Frege successfully proved a categoricity theorem like Dedekind's. And it is incorrect systematically because essentially the same argument establishes the categoricity of second-order arithmetic in any of the usual systems of set theory. And it is surely implausible that only someone familiar with the categoricity of the Peano–Dedekind axioms as a theorem of second-order logic has really grasped the theorem or its proof. At most, Frege might be charged with having missed a subtlety concerning the distinction between formal and semi-formal systems; but this is hardly surprising for the period in which he wrote.

Bell and I accepted Dummett's view in *FPL* and based our claim that Frege's interpretation of the function variables was the standard one on the premise that Frege's concept of a function coincides with the set-theoretic notion of an arbitrary correspondence. In this case the domain of the function variables is in one–one correspondence with the power set of the domain of the individual variables. Our thought was that whatever covert role the neglect of Cantor's theorem might have played in the inconsistency of *Grundgesetze*, it is unlikely that Frege sought to ignore the theorem by assuming that the totality of functions, like the totality of expressions, is countably infinite. But we sided with the view Dummett expressed in *FPM* and supposed that Frege might very well have been misled into assuming that what holds for certain countable interpretations of the function variables holds in general; hence we agreed with Dummett's evaluation of the sense in which Frege missed the significance of the possibility of alternative interpretations for his program.

More recently, reflection occasioned by reading Sandu's (2005) contribution to the Frapolli (2005) collection of essays on Ramsey has convinced me that the equation of Frege's concept of a function with the notion of an arbitrary correspondence should be reconsidered, and that it might be fruitful to reconsider it from the perspective of Ramsey's interpretation of *Principia*'s propositional functions.

Let us call *classes* the extensions of functions in the *logical* sense of 'function,' a notion I will explain in the sequel. Then we wish to explore whether the classes determined by such functions correspond to only a fragment of the sets associated with arbitrary correspondences. The assumption that something like this is correct evidently underlies Ramsey's proposed reinterpretation of the first-edition *Principia*'s notion of a propositional function in his essay, *The Foundations of Mathematics*. That essay is usually cited for its separation of the paradoxes and its isolation of the simple hierarchy of *Principia*'s functions of lowest order. It is generally assumed that by this hierarchy Ramsey understood a hierarchy of functions interchangeable with the standard hierarchy of sets, modulo the condition that *Principia*'s hierarchy is stratified rather than cumulative. In fact, Ramsey regards *Principia*'s simple hierarchy of functions of lowest order as inadequate, and he devotes a central chapter of his essay (Chapter IV) to criticizing *Principia*'s notion of propositional function and arguing for the notion's extension. He achieves this extension, and more, by the introduction of the notion of a propositional function in extension, or what I will call an *extensional propositional function*. The consideration of Ramsey's criticisms of *Principia* has led me to the view that the interest of Hintikka and Sandu's paper has less to do with standard versus nonstandard interpretations of second-order logic than with Frege's concept of a function. As I now see it, the chief interest of their paper, although not perhaps their principal aim, is the suggestion that the difference between logical and set-theoretic notions of *function* parallels the well-known contrast between the logical notion of *class* and the mathematical notion of *set*.

My strategy for the balance of the discussion is to proceed antihistorically by reviewing Ramsey's notion of an extensional propositional function, its difference from what Ramsey calls the *predicative* propositional functions of *Principia*, and the uses to which Ramsey put the notion in his defense of a modified logicist position. To motivate the notion of an extensional propositional function, I will begin by briefly recounting the relevant Tractarian background to Ramsey's thought. I will then describe two reconstructions of the Axiom of Infinity, one with, and one without the notion of an extensional propositional function. Both reconstructions trace back to Ramsey. The reconstruction involving the notion of an extensional propositional function was elaborated in his *FoM*; my discussion of it constitutes the principal novelty of the present chapter. Having clarified the predicative and extensional notions, I will return to Frege's functions. I will argue that they have a feature in common with *Principia*'s propositional functions, and that it is plausible to argue, on the basis of this point of similarity, that they should be distinguished from the notion of an arbitrary correspondence. However, the situation is not entirely straightforward, and we will see that there are also reasons to suppose that Frege's notion can be regarded as falling in line with the mathematical (or "extensionalist") tradition.

II.2 THE TRACTARIAN BACKGROUND TO RAMSEY'S EXTENSIONAL PROPOSITIONAL FUNCTIONS

There are three key ideas of the *Tractatus* that form the background to Ramsey's notion of an extensional propositional function: (i) a proposition is significant only insofar as it partitions all possible states of affairs into two classes. This is what Michael Potter[7] has called "Wittgenstein's big idea." It implies that tautologies cannot be regarded as significant propositions, and in so far as the propositions of logic are all tautologies, the propositions of logic are not significant propositions; in particular, they cannot be merely more general than the truths of zoology or any other special science as Russell once claimed (Russell 1919, p. 169). (ii) Objects, which constitute the substance of the world, are constant across alternative possibilities; so also therefore is their number. Potter puts the point well when he says that "what changes in the transition between [possibilities] is how objects are combined with one another to form atomic facts; what the objects are does not change because they are the hinges about which the possibilities turn and hence are constant" (2005, p. 72). It follows that there cannot be a significant proposition regarding the number of objects. At best language

---

[7] In Potter (2005). My exposition of the Tractarian background to Ramsey's thought is indebted to this paper by Potter and also to Potter (2000).

can "reflect" the cardinality of the world. (iii) Since atomic propositions are logically independent of one another, statements of identity cannot express significant propositions, for if they did, they would establish relations of logical dependence among atomic propositions.

It is a consequence of Wittgenstein's rejection of identity, a rejection with which Ramsey concurred, that the notion of class implicit in Russell's theory of classes must be an "accidental" one, that is, one according to which it is possible that every property might be shared by at least two individuals, with the consequence that there would be nothing to answer to the number 1.[8] If we could express significant propositions with the use of identity, then among the properties of *a* there would be the property of *being identical with a*, and this would settle the question of the existence of the number 1 on Russell's theory. But acceptance of the third Tractarian thesis argues against a solution which, like this one, treats identity as a possible constituent of a significant proposition. Hence if we follow the *Tractatus* on identity, Russell's theory cannot recover the notion of class that its account of number – and indeed, of the whole of mathematics – requires. This is an objection that is in some ways more fundamental than the standard objection to the Axiom of Infinity, since even if that axiom could in some way be justified, according to the present objection, it is still *possible* on Russell's theory that there might not be enough numbers to go around. Hence, for this and other reasons we will soon come to, mathematics cannot be based on *Principia*'s theory of classes.

Ramsey's notion of an extensional propositional function emerged from his attempt to address this and other defects in *Principia*'s theory of classes, such as those surrounding its account of the Axiom of Choice. But perhaps the most important consideration in favor of the notion was that it proved essential to a formulation of Infinity that accords with the first and second Tractarian theses. For, if we follow Russell and simply take Infinity to assert that for every inductive cardinal number *n*, there is a class whose cardinality is given by *n*, then we treat the axiom as an accidental truth concerning the number of things of a particular sort. As a consequence, we miss the peculiar character of claims regarding the cardinality of objects – in the present case, of individuals – enunciated by (ii): the existence of objects is a precondition for significant propositions and cannot be a subject with which such propositions deal; the same is true of their number. In order to see how Ramsey's notion of an extensional propositional function leads to a

---

[8] *FoM*, p. 50/213; my references to Ramsey's *FoM* give the page number of the Braithwaite (1931) edition of his papers followed by the page number of the Mellor (1990) edition.

formulation of Infinity that addresses these issues, it will be useful to begin with a formulation of the axiom that has all the ingredients of the formulation of *FoM except* for the notion of an extensional function.

Potter (2005) discusses an early unpublished fragment of Ramsey's (entitled "The number of things in the world"), which contains what Potter calls "Ramsey's transcendental argument for Infinity." Potter canvasses a number of questions of Ramsey-interpretation which the fragment raises. The transcendental argument evidently precedes Ramsey's discovery of extensional propositional functions. For my purposes, the argument's most interesting feature is the formulation of Infinity it suggests, rather than the considerations in favor of the axiom it advances. I will refer to this as "Ramsey's early formulation of Infinity"; however, my discussion is not based on Ramsey's unpublished fragment, but on Potter's reconstruction of it. The historical accuracy of Potter's account is of course completely irrelevant to its usefulness for motivating the formulation of Infinity that Ramsey gives in *FoM*. That formulation is my main concern, and I will take it up in the next section.

Let φx be any propositional function, and let Tx be $\varphi x \vee \neg \varphi x$. Then to say that there are at least n individuals write

$$p_n =_{Df} \exists x_1 \ldots \exists x_n \, Tx_1 \wedge \ldots \wedge Tx_n,$$

where, here and elsewhere, Ramsey assumes Wittgenstein's "nested variable convention," which says that whenever a variable occurs within the scope of another variable, it is to be assigned a different individual as its value.[9] Thus, for example, '$\exists x \exists y \ldots$' is read, 'there is an individual x and *another* individual $y \ldots$' Then if there are n individuals $a_1, \ldots, a_n$, this is reflected by the tautology $Ta_1 \wedge \ldots \wedge Ta_n$.

Now let p_ω be the "logical product" of the p_n, for all finite n. Then p_ω is Ramsey's early formulation of Infinity, where what Ramsey means by the logical product of the p_n can be gleaned from what he says about logical sums:

A logical sum is not like an algebraic sum; only a finite number of terms can have an algebraic sum, for an "infinite sum" is really a limit. But the logical sum of a set of

---

[9] What I am calling Wittgenstein's nested variable convention is explained in detail in Wehmeier (2008) in his discussion of "W-logic." Wehmeier (2009) discusses a variant of W-logic ("R-logic") which originated from Ramsey's initial misunderstanding of Wittgenstein's convention.

propositions is the proposition that these are not all false, and exists whether the set be finite or infinite. (*FoM*, p. 56/219, footnote 1)

Ramsey clearly takes this to apply mutatis mutandis to the notion of a logical product, in which case we are to understand by p_ω the proposition that every member of the p-series is true. This is expressed by the infinite conjunction of the p_n, which is what Ramsey means by the proposition that every member of the p-series is true. Hence, if there are infinitely many individuals, this is reflected by the tautology $Ta_1 \wedge \ldots \wedge Ta_n \wedge \ldots$, which asserts, or appears to assert, the infinity of the number of objects of the type of individuals, and is to be contrasted with a claim about the number of things there are of some particular kind.

To construct an "Axiom of Infinity" regarding some sort of thing, we also proceed by first describing a series of propositions such as

$q_1 =_{Df} \exists x(x$ is a hydrogen atom$)$
$q_2 =_{Df} \exists x \exists y(x$ and y are hydrogen atoms$)$
etc.

where once again Wittgenstein's convention regarding nested variables is assumed. In this case, each q_n is an ordinary empirical proposition concerning not objects in general but things of a particular kind, and q_ω (the product of all the q_n) says that there are infinitely many of them. The *Tractatus* imposes no prohibition against the significance of a proposition like q_n since it is a genuine possibility that there are at least n hydrogen atoms, even if there are in fact only m of them for m less than n. It is likewise a genuine possibility that there are *infinitely many* hydrogen atoms. All such claims mark genuine possibilities in the sense demanded by the Tractarian notion of a significant proposition.

There is a well-known distinction of Carnap's that bears on the evaluation of Ramsey's early proposal. Carnap (1931, p. 41) distinguished between two types of logicist reduction. Let us call a reduction *type (a)* if it defines all concepts of a mathematical theory in terms of those of logic, and *type (b)* if, in addition to providing definitions of a mathematical theory's concepts in terms of logical concepts, it derives the axioms of the theory from purely logical axioms. It is worth remarking that although Russell may have sometimes expressed the view that a type (a) reduction suffices for logicism, Frege was always clear on the need to secure both type (a) and type (b) reductions.

In *FoM* (pp. 3–4/166–7), which was written after the fragment containing Ramsey's early formulation of Infinity, there is an anticipation of Carnap's distinction between type (a) and type (b) reductions. Ramsey gave the distinction an interesting interpretation when he remarked – clearly under the influence of Wittgenstein – that while a type (a) reduction might illuminate the *generality* of mathematics, a type (b) reduction would address what is truly distinctive about it: its *necessity*. But if Ramsey and Wittgenstein are the source of the view that necessity is the chief characteristic of mathematical propositions that requires explanation, it is important to remember that there is little interest in necessity in Frege's *Grundlagen* or in the first edition of *Principia*, and no interest in anything that could be called "metaphysical necessity." Both works stress the generality of mathematics and the possibility that this might be illuminated by assimilating it to the generality of logic. Ramsey's aim was to illuminate the necessity of mathematics by assimilating it to the necessity of logic. But he also saw that there are at least two ways in which this might be accomplished. We can try reducing the propositions of mathematics to those of logic after the fashion of a type (b) reduction of the sort Frege tried, unsuccessfully, to achieve. Alternatively, we might attempt to show that the propositions of mathematics have the same kind of necessity that we find in the propositions of logic – not by deriving mathematics from logic, but by an analysis of the necessity mathematical propositions themselves exhibit. Ramsey sought to explicate this characteristic by proposing that the propositions of mathematics, like those of logic, are "tautologies," an idea that evidently has its basis in the *Tractatus*. However, by a tautology Ramsey did not mean something that is *truth-table decidable*, as is clear from his discussion of the Axiom of Choice where he mentions with approval the possibility of "a tautology, which could be stated in finite terms, whose proof was, nevertheless, infinitely complicated and therefore impossible for us" (*FoM*, p. 59/ 222). We will return to Ramsey's notion of *tautology*.

How do these considerations bear on Ramsey's early formulation of Infinity? There are two objections to this formulation, one more telling than the other. The less telling objection is that the existential claims expressed by the propositions of the *p*-series fall short of possessing the necessity of logical propositions. The fact that $Tx_1 \wedge \ldots \wedge Tx_n$ becomes a tautology when there are individuals $a_1, \ldots, a_n$ which accord with Wittgenstein's convention does not show that p_n is a tautology, and therefore does not show that the axiom is necessary. In a domain of $m < n$ objects, p_n is simply false, a fact which on Ramsey's formulation is represented by the absence of values for the variables of p_n that accord with Wittgenstein's nested variable convention. But to this objection Ramsey can respond that it was not his intention to show that p_n, let alone p_ω, is a tautology. Rather, the point of the formulation was to capture the idea that if there are n individuals, then p_n is *witnessed* by the tautology, $Ta_1 \wedge \ldots \wedge Ta_n$. The objection mistakenly assumes that the only way to establish the logical necessity of Infinity is to show that it holds in every "universe of discourse." This is one sense in which a proposition may be seen to be a logical proposition, but it is not the sense Ramsey's formulation was intended to capture. A proposition can be one whose truth is witnessed by a tautology, and thus be a logical proposition in Ramsey's sense, without holding in every universe of discourse. And this is precisely the case with Infinity.

The second objection is harder to articulate and will only become clear after we have examined the formulation of Infinity in *FoM* that uses Ramsey's notion of an extensional propositional function. As a first approximation, the difficulty is that the tautological character of the propositional function Tx which figures in Ramsey's *p*-series is not integrated into the theory of functions and classes as it would be on a formulation of Infinity that was derived from an analysis of *class* and *propositional function*. For this reason, Ramsey's formulation carries little conviction as an account of the necessity that a fundamental axiom like Infinity is supposed to exhibit. Since, according to Ramsey's reading of the *Tractatus*, the explanation of this characteristic is the central task of a philosophy of mathematics, it seems likely that the source of his dissatisfaction with his early proposal (and perhaps the reason he never published it) was the recognition that it failed to present a convincing account of Infinity's necessity.

II.4 INFINITY WITH EXTENSIONAL PROPOSITIONAL FUNCTIONS

In *FoM* Ramsey rejected the notion of propositional function that we associate with the 1910 *Principia*. Such functions are, in Ramsey's phrase, "predicative" in the sense that the proposition φa which the propositional function φ assigns to a says or *predicates* the same thing of a as the proposition φb, which φ assigns to b, does of b. This connection with predication is essential to any *logical* notion of the functions which determine classes, and it stands in contrast with the idea that a function is an *arbitrary* correspondence. According to Ramsey's new conception, the way to accommodate the broader notion of set which is implicit in the idea of an arbitrary correspondence is by conceiving of a propositional function as an arbitrary mapping, in the present case from individuals to propositions, with φa and φb merely the values of φ for a and b as arguments. Here 'arbitrary' means both that the mapping is not constrained to preserve a property in the sense that the propositions φa and φb are not required to say the same thing of *a* and *b*, *and* that all combinatorial possibilities of functional pairings of individuals with propositions are allowed. On Ramsey's view, it is only *propositions* that involve properties which are predicable of individuals; the *classes* which propositional functions determine should not be constrained by the demand that their elements share a common property.[10]

When this purely extensional notion of a propositional function is adopted, there is, of course, no difficulty with the existence of unit classes and the number 1, since every object has a mapping to propositions that is unique to that object.[11] Notice, however, that, of the two features of extensional propositional functions, it is only the second, namely the fact that such functions exhaust all combinatorial possibilities, that the solution to this difficulty depends upon, and this can readily be accommodated within a framework that considers only pairings of individuals with truth values. The "propositional" aspect of propositional functions, the fact that they map individuals to propositions, does not enter in any essential way into the solution of the problem raised by the possibility that every property is shared by at least two individuals.

Ramsey says of the Axiom of Choice that, given his notion of an extensional propositional function, "it is the most evident tautology ... [and not something that] can be the subject of reasonable doubt" (*FoM*, p. 58/221). He then proceeds to show how, on Whitehead and Russell's understanding of *propositional function* and *class*, the axiom, though not a contradiction, is also not a tautology. Ramsey does not pause to explain why on his account the axiom is an evident tautology, but as John Bell once observed in conversation, after the existence of singletons and sums has been shown to follow from the concept of an extensional propositional function, Choice, in the form 'If *K* is a family of disjoint nonempty classes, then *K* has a choice class,' is clearly true. For if classes are determined by extensional functions, it is evident that every class in *K* can be shrunk to a singleton; the sum of all such singletons is the

---

[10] By an abuse of notation which uses the same symbol φ in two very different ways – both for a mapping from individuals to propositions and for a predicate of the language – a predicative propositional function φ takes an individual *a* to a proposition of the form φ*a*, i.e. to a proposition which is expressed by a sentence consisting of the concatenation of the predicate φ with the constant *a*. The class determined by φ is the class of all *a* such that φ*a* is true. If φ is extensional but not predicative, the class determined by φ is the class of all individuals *a* which φ maps to truths. Under an extensional understanding of propositional functions, there is not in general a correspondence between propositional functions and predicates of the language, so that the association with propositions is in this sense "arbitrary." It is also arbitrary in the stronger sense of allowing all possible pairings of individuals with truth values.

[11] For example the singleton of *a* is determined by a propositional function which maps *a* to an arbitrarily selected truth and maps every other individual to a falsehood.

required choice class. Like the difficulty with the number 1, the foregoing justification of Choice depends only on the fact that extensional propositional functions exhaust all combinatorial possibilities.

Clearly, what Ramsey means by the tautologousness of Choice is not captured by the idea that it holds in all universes of discourse, still less that it is truth-table decidable. Ramsey perceived that there are "interpretations" of *Principia*, by which he meant understandings of the notion of *propositional function* – of which the one favored by Whitehead and Russell is just one example – under which Choice can be shown to be false. And in a remark made in the course of a discussion of Infinity and the possibility of saying something about the cardinality of the world given his adherence to the second Tractarian idea to which we earlier called attention in Section 11.2, Ramsey shows his appreciation of the possibility of falsifying a fundamental axiom by "imagining a universe of discourse, to which we may be confined, so that by 'all' we mean all in the universe of discourse" (*FoM*, p. 61/224). From this we may conclude that Ramsey's understanding of the "tautologousness" of Choice is that relative to his extensional understanding of 'propositional function' and the intended meaning of 'all,' and relative perhaps as well to the intended meanings of the propositional connectives, Choice is evidently true.[12] Ramsey expresses this by saying that, under these conditions, the axiom is "an evident tautology," to draw attention to the fact that its obvious demonstration in the finite case proceeds by inspecting all the combinatorially possible relations between classes and their elements. In the general case, we are incapable of carrying out such an inspection, but this does not affect the truth of the principle – assuming an understanding of *propositional function* that admits all combinatorial possibilities.

The situation is different with Ramsey's analysis of the Axiom of Infinity. Here, in addition to the fact that Ramsey's functions exhaust all combinatorial possibilities, the *propositional* character of extensional propositional functions is an indispensable component of the solution to the problem the axiom presents. To understand Ramsey's account of Infinity, suppose we try expressing the idea that there are at least two things by the proposition

$$(^*)\exists x\exists y\neg(\varphi)(\varphi x \equiv \varphi y),$$

---

[12] Sandu misses this point about the tautologousness of Choice when he criticizes Ramsey's contention that Choice is a tautology and argues that since "there are models of set theory in which the Axiom of Choice is false ... it cannot, therefore, be a tautology" (Sandu 2005, p. 252).

where it is assumed that propositional functions are to be understood predicatively, after the manner of Whitehead and Russell. Now consider a universe of discourse U containing precisely two individuals a and b which have all their properties in common. For Ramsey there is nothing absurd in the idea that two things might share all their properties and thus be indistinguishable but distinct; and since he rejects identity he will not allow an appeal to the property, *being identical with a*, for example, in order to ensure the truth of (*) in U under a predicative understanding of propositional functions. In these circumstances, (*) will fail to reflect the fact that a and b comprise a two-element universe. But for Ramsey that it should even be *possible* that (*) can fail in this way shows that, under its predicative interpretation, (*) purports to express a general truth, and hence a significant proposition; it cannot therefore be the correct expression of the idea that there are at least two things.

Let us now consider what happens when propositional functions are understood extensionally. Assuming Wittgenstein's nested variable convention, (*) *must* be true in any two-element universe such as U. For if we consider all possible mappings φ, there must be one among them that assigns a to the negation of whatever proposition it assigns b. Hence the function $\neg(\varphi)(\varphi x \equiv \varphi y)$ will map (a, b) to the negation of a contradiction, and therefore on Ramsey's understanding of *propositional function*, the truth of (*) in U will be witnessed by a tautology. This contrasts with the predicative interpretation under which (*) can fail in a two-element universe, so that even in cases where it holds, it does so under a predicative understanding of *propositional function* only in virtue of a contingent fact about individuals and their properties. It also contrasts with Potter's reconstruction of Ramsey's early formulation, since the fact that (*) is witnessed by a tautology is not an ad hoc stipulation, but a consequence of Ramsey's extensional understanding of the nature of a propositional function.

In light of the foregoing considerations, let us define the propositional function

$$T(x, y) =_{Df} (\varphi)(\varphi x \equiv \varphi y).$$

As we have seen, when propositional functions are understood extensionally, the function $T(x, y)$ maps to a tautology when the values of x and y are the same and to a contradiction otherwise. Thus if propositional functions are understood extensionally, then if there are two individuals, and we understand $\exists x \exists y T(x, y)$ to say, 'There is an x and *another* y such that

$T(x, y)$,' then $\exists x \exists y T(x, y)$ reduces to a contradiction, while $\exists x \exists y \neg T(x, y)$ reduces to a tautology.[13] This suggests a *new p*-series defined as follows:

$$p_1 =_{Df} \exists x T(x, x)$$

$$p_2 =_{Df} \exists x \exists y \neg T(x, y)$$

$$p_3 =_{Df} \exists x \exists y \exists z \neg T(x, y) \wedge \neg T(x, z) \wedge \neg T(y, z)$$

etc.

For each n, p_n is true in a universe of n individuals. The witnesses to the propositions of the series alternate between a tautology, when the member of the series is true, and a contradiction, when it is false. Ramsey's new formulation of the Axiom of Infinity is given by the logical product $p_{\aleph_0}$ of the propositions of this series. Observe that if there are $\aleph_0$ individuals, $p_{\aleph_0}$ is witnessed by a product of tautologies, and if fewer than $\aleph_0$, its falsity is witnessed by a contradiction.

To appreciate Ramsey's achievement, consider how his proposal differs from the following reduction of truths to tautologies. (For simplicity, we restrict ourselves to quantifier-free molecular formulas, since this suffices to illustrate the point of difference with the alternative reduction to which I wish to call attention.) Replace a quantifier-free molecular formula φ by its equivalent disjunctive normal form. Next replace any true literal[14] of φ's disjunctive normal form by $x = x$, and replace any false literal by $x \neq x$. Then the resulting formula is a truth function of "tautologies" and "contradictions," and it reduces to one or the other according to whether it is true or false. This is clearly artificial since it enables us to "reduce" to tautologies and contradictions many propositions which are merely true or false. But *FoM*'s proposal regarding Infinity does not simply replace truths with tautologies, and falsehoods with contradictions; it *derives* the tautologous or contradictory character of a witness to a proposition of the *p*-series from an analysis of the notion of a propositional function: a proposition of the *p*-series is witnessed by a tautology or contradiction as a consequence of the cardinality of the domain *and the extensional character of propositional*

---

[13] As we noted earlier, under a predicative interpretation we can read off the proposition to which φ maps a from the propositional function-expression 'φa'; its sentential expression consists of the predicate φ followed by the constant a. In particular, T maps (a, b) to the proposition $(\varphi)(\varphi a \equiv \varphi b)$, which has a natural reading given the connection between predicative functions and predicates. Under Ramsey's extensional interpretation, this connection is severed, and $(\varphi)(\varphi a \equiv \varphi b)$ is the infinite conjunction of propositions of the form $p \equiv q$, where p and q are (respectively) the values of some φ for a and b as arguments.

[14] A *literal* is an atomic formula or the negation of an atomic formula.

functions. In this respect Ramsey's reduction procedure stands in marked contrast with one which merely stipulates the tautologousness of the witness to the truth of a proposition.

Does the formulation of Infinity based on Ramsey's reinterpretation of the notion of a propositional function provide a logical justification of Infinity in the usual sense? It does not. Ramsey's achievement consists in showing how the witness to the truth of Infinity reduces to a product of tautologies when the cardinality of the domain of individuals is infinite. In conformity with the first of the Tractarian theses noted earlier, the truth of the axiom is not expressed by a genuine proposition; rather, the cardinality of the world is "shown" – or, as I prefer to say, "witnessed" – by a tautology. This is not the provision of a logical justification for Infinity that purports to show that it is true in every universe of discourse, but an explanation of how, *as a consequence of the notion of an extensional propositional function*, a witness to the truth of the axiom reduces to a product of tautologies in any domain in which the axiom holds, and how a witness to its falsity reduces to a contradiction in any in which it does not hold. It therefore addresses the second of the two objections we noted in our discussion of the formulation of the unpublished fragment at the conclusion of Section 11.3.

11.5 EXTENSIONAL PROPOSITIONAL FUNCTIONS AND LOGICISM

Hintikka and Sandu's discussion of Fregean concepts shares a number of similarities with Ramsey's discussion of the 1910 *Principia*'s propositional functions: like Ramsey, Hintikka and Sandu regard the notion of *class* on which the early logicists relied as inadequate by comparison with the extensionalist tradition's notion of *set*; and both Hintikka and Sandu, and Ramsey base their analyses of the inadequacy of classes on a failure of the early logicists to conceive of a function as an arbitrary correspondence. But there are also significant differences. By contrast with Hintikka and Sandu, Ramsey took himself to be contributing to a more defensible version of logicism. He hoped to show that his formulation of an appropriate extensionalist replacement of *Principia*'s predicative propositional functions would address an internal difficulty with the work, one surrounding the adequacy of its account of mathematical propositions.

Ramsey's extensional propositional functions attempt to marry two seemingly incompatible ideas: extensional propositional functions preserve the letter of the logicist thesis that classes are determined by functions, but they also invoke the combinatorial notion of set by representing a propositional function as an arbitrary many–one pairing of individuals with propositions. As a consequence, one gives up the Russellian idea that a propositional function determines a class in terms of an antecedently available property. But Ramsey's extensionalism is in some respects congenial to Frege. This is because Frege's functions satisfy a condition of extensionality in the sense that functions whose courses of values coincide are not distinguished, and because Fregean functions form a simple hierarchy which is not constrained by ramification. Despite these points of agreement with an extensionalist view-point, Frege's assimilation of concepts to functions which map into truth values is usually understood to share with the 1910 *Principia* the idea that the correspondence is not arbitrary, but is constrained by the principle that if a function maps two objects to The True, they must fall under a common concept. Hence Frege's generalization of the function concept to include those that map to truth values is, in Ramsey's sense of the term, a generalization to a notion of function that is just as predicative as Russell's.

If I have understood him, Sandu's (2005) suggestion is that his paper with Hintikka should be understood as arguing that the predicativeness of Frege's functions is sufficient to show that they do not exhaust all mappings between objects and truth values, and that this lends support to his and Hintikka's claim that *in general* Frege's functions are not arbitrary correspondences. But it is at this point that I think a difference between *Principia*'s propositional functions and Frege's concepts may be important. For *Principia*, a function maps to the truth values only by "passing through" a proposition; this, after all, is why they are called "*propositional* functions." But Frege's concepts map "directly" to the truth values. To be sure, the informal explanation of the Fregean idea that concepts are a kind of function invariably proceeds by saying that a function maps an object to The True just in case the object falls under the associated concept. But as pedagogically natural as this explanation may be, it is also misleading, since Frege's assimilation of concepts to functions does not exclude the possibility that a concept might be no more accessible to us than the correspondence which a mapping to truth values establishes. In such a case, there would be little to choose between a Fregean function and the extensional notion of a function as an arbitrary correspondence between objects and truth values.

What of Hintikka and Sandu's question: What was Frege's concept of *function*? Frege explains our acquisition of the functions associated with concepts – functions which express a judgeable content – by our acquisition of their linguistic expression, something we achieve by an analysis of sentences. For Frege this may well have been an essential component of his conception of functions and concepts. But like our informal explanation of how to understand functions which map to truth values, perhaps it, and the predicative interpretation it suggests, can be regarded as a consideration whose purpose is only to motivate the notion. The question is whether, by allowing for functions as arbitrary correspondences between objects and truth values, we do as much violence to the Fregean idea of a concept or function as Ramsey's extensional functions do to the propositional functions of *Principia*. As often happens, settling on an answer to an interpretive question such as this one is less important than achieving greater clarity on the systematic issues the question raises.

It seems clear that both *Principia*'s propositional functions and Frege's concepts admit of extensionalist interpretations – in the strong sense considered here, of arbitrary mappings between *Principia*'s individuals and propositions, or between Frege's objects and truth values. Such interpretations certainly tend to undermine the motivation for both ideas, a fact that is particularly evident in the case of Russell's propositional functions, where the extensionalist interpretation seems more properly regarded as a wholesale *replacement* of the notion. The situation is less clear in the case of Fregean functions and concepts because they lack the explicit association with propositions that is characteristic of propositional functions; an extensionalist interpretation of a Fregean concept as an arbitrary mapping of objects to truth values is arguably still a Fregean concept. However, its utility for Frege's theory of classes is unclear. According to a theory like Frege's, concepts provide the principle which gives classes their "unity," and they also serve the epistemological function of providing the principle under which a collection of objects can be regarded as a separate object of thought. A class that is generated by an arbitrary pairing of individuals with truth values might be one that is "determined by a concept," but the concept which determines it seems no more epistemically accessible than the collection itself. Even if it can be convincingly argued that such concepts sustain the "unity" of the classes they determine, it can hardly be maintained that they are capable of playing the epistemological role which the predicative interpretation can claim for its functions and concepts.

ACKNOWLEDGMENTS

I wish to thank Gerry Callaghan, Greg Lavers, and Peter Koellner for their insightful comments on earlier drafts. I am indebted to Michael Hallett for suggestions which led to many improvements. I owe a special acknowledgment to John Bell for many hours pleasantly spent discussing the issues dealt with here. Support from the Killam Trusts and the Social Sciences and Humanities Research Council of Canada is gratefully acknowledged.

Bibliography

Ainsworth, P. M. (2009). "Newman's objection," *British Journal for the Philosophy of Science* 60: 1–37.

Anderson, C. A. (1989). "Russell on order in time," in Savage and Anderson (1989: 249–63).

Beck, L. W. (1950). "Constructions and inferred entities," *Philosophy of Science* 17: 74–86.

Bell, D. (1990). "How 'Russellian' was Frege?," *Mind* 99: 267–77.

Bell, J. L. (1995). "Type-reducing correspondences and well-orderings: Frege's and Zermelo's constructions re-examined," *Journal of Symbolic Logic* 60: 209–21.

(1999a). "Frege's theorem in a constructive setting," *Journal of Symbolic Logic* 64: 486–8.

(1999b). "Finite sets and Frege structures," *Journal of Symbolic Logic* 64: 1552–6.

Benacerraf, P. (1981). "Frege: the last logicist," in P. French *et al.* (eds.), *Midwest Studies in Philosophy*, vol. VI (Minneapolis: University of Minnesota Press): 17–35, reprinted in Demopoulos (1995: 41–67).

Bernays, P. (1935). "Sur le platonisme dans les mathématiques," *L'Enseignement Mathématique* 34: 52–69, trans. C. Parsons as "On Platonism in mathematics," reprinted in Putnam and Benacerraf (1983: 258–71).

Boolos, G. (1985). "Reading the *Begriffsschrift*," *Mind* 94: 331–44, reprinted in Demopoulos (1995: 163–81) and in Boolos (1998: 155–70).

(1987). "The consistency of Frege's *Foundations of Arithmetic*," reprinted in Demopoulos (1995: 211–33) and in Boolos (1998: 183–201).

(1990). "The standard of equality of numbers," reprinted in Demopoulos (1995: 234–56) and in Boolos (1998: 202–19).

(1994). "The advantages of honest toil over theft," reprinted in Boolos (1998: 255–74).

(1998). *Logic, Logic, and Logic* (Cambridge, MA: Harvard University Press).

Bottazzini, U. (1986). *The Higher Calculus: A History of Real and Complex Analysis from Euler to Weierstrass* (Berlin and New York: Springer-Verlag).

Braithwaite, R. B. (1953). *Scientific Explanation: A Study of the Function of Theory, Probability and Law in Science* (Cambridge University Press).

(ed.) (1931). *The Foundations of Mathematics and other Logical Essays by Frank Plumpton Ramsey* (Paterson, NJ: Littlefield and Adams).

Burge, T. (2005). "Frege on apriority (2000)," in T. Burge, *Truth, Thought, Reason: Essays on Frege* (Oxford University Press): 356–89.

Burgess, J. P. (1993). "Hintikka *et* Sandu *versus* Frege *in re* arbitrary functions," *Philosophia Mathematica* series III, 1: 50–65.

(2005). *Fixing Frege* (Princeton University Press).

Carnap, R. (1928). *Der logisches Aufbau der Welt*, trans. R. A. George as *The Logical Structure of the World* (University of California Press, 1967).

(1931). "The logicist foundations of mathematics," trans. E. Putnam and G. Massey, reprinted in Putnam and Benacerraf (1983: 41–51).

(1934). *Logical Syntax of Language*, trans. A. Smeaton (London: Routledge and Kegan Paul, 1937).

(1939). *Foundations of Logic and Mathematics*, in *International Encyclopedia of Unified Science*, vol. 1, part 1 (University of Chicago Press): 139–213.

(1942). *Introduction to Semantics* (Cambridge, MA: Harvard University Press).

(1950). "Empiricism, semantics and ontology," *Revue Internationale de Philosophie* 4: 20–40, revised and reprinted in R. Carnap, *Meaning and Necessity: A Study in Semantics and Modal Logic* (University of Chicago Press, 1956, enlarged edn): 204–21. Page references are to the 1956 reprinting.

(1963a). "Intellectual autobiography," in Schilpp (1963: 3–84).

(1963b). "Replies and systematic expositions," in Schilpp (1963: 859–1013).

(1966). *Philosophical Foundations of Physics: An Introduction to the Philosophy of Science*, ed. Martin Gardner (New York: Basic Books).

(1995). *An Introduction to the Philosophy of Science*, ed. M. Gardner (New York: Dover), reprint of the second (1974) edn of Carnap (1966).

Cartwright, R. (1987). "On the origin of Russell's theory of descriptions," in R. Cartwright, *Philosophical Papers* (Cambridge, MA: MIT Press): 95–134.

Carus, A. (2007). *Carnap and Twentieth-Century Thought: Explication as Enlightenment* (Cambridge University Press).

Chalmers, A. (2011). "Drawing philosophical lessons from Perrin's experiments on Brownian motion: a response to van Fraassen," *British Journal for the Philosophy of Science* 62: 711–32.

Chambers, T. (2000). "A quick reply to Putnam's paradox," *Mind* 109: 195–7.

Church, A. (1984). "Russell's theory of identity of propositions," *Philosophia Naturalis* 21: 513–22.

Coffa, A. (1982). "Kant, Bolzano and the emergence of logicism," *Journal of Philosophy* 74: 679–89, reprinted in Demopoulos (1995: 29–40).

Craig, W. and R. L. Vaught (1958). "Finite axiomatizability using additional predicates," *Journal of Symbolic Logic* 23: 289–308.

Dedekind, R. (1872). "Continuity and irrational numbers," in Dedekind (1963: 1–27).

(1888). "The nature and meaning of numbers," in Dedekind (1963: 31–115).

(1890). "Letter to Keferstein," trans. S. Bauer-Mengelberg and H. Wang, reprinted in van Heijenoort (1967: 98–103).

(1963). *Essays on the Theory of Numbers*, trans. W. W. Beman (New York: Dover).

Demopoulos, W. (1994). "Frege and the rigorization of analysis," *Journal of Philosophical Logic* 23: 225–45, reprinted in Demopoulos (1995: 68–88).

(1999). "On the theory of meaning of 'On denoting,'" *Noûs* 33: 439–58.

(2003a). "On the rational reconstruction of our theoretical knowledge," *British Journal for the Philosophy of Science* 54: 371–403.

(2003b). "Russell's structuralism and the absolute description of the world," in N. Griffin (ed.), *The Cambridge Companion to Russell* (Cambridge University Press): 392–419.

(2007). "Carnap on the rational reconstruction of scientific theories," in Friedman and Creath (2007: 248–72).

(2011). "On logicist conceptions of functions and classes," in D. DeVidi, M. Hallett, and P. J. Clark (eds.), *Vintage Enthusiasms: Essays in Honour of John L. Bell* (Dordrecht: Springer): 3–18.

(2011). "Three views of theoretical knowledge," *British Journal for the Philosophy of Science* 62: 177–205.

(2011). "On extending 'Empiricism, semantics and ontology' to the realism-instrumentalism controversy," *Journal of Philosophy* 108: 647-69.

(ed.) (1995). *Frege's Philosophy of Mathematics* (Cambridge, MA: Harvard University Press).

Demopoulos, W. and J. L. Bell (1993). "Frege's theory of concepts and objects and the interpretation of second-order logic," *Philosophia Mathematica* series III, 1: 139–56.

Demopoulos, W. and P. J. Clark (2005). "The logicism of Frege, Dedekind, and Russell," in S. Shapiro (ed.), *The Oxford Handbook of Philosophy of Mathematics and Logic* (Oxford University Press): 129–65.

Demopoulos, W., M. Frappier, and J. Bub (2012). "Poincaré's 'Les conceptions nouvelle de la matière'," *Studies in History and Philosophy of Modern Physics* 43: 221–5.

Demopoulos, W. and M. Friedman (1985). "Bertrand Russell's *The Analysis of Matter*: its historical context and contemporary interest," *Philosophy of Science* 52: 621–39.

DePierris, G. (2012). "Hume on space, geometry, and diagrammatic reasoning," *Synthese* 186: 169–90.

DeVidi, D. and G. Solomon (1994). "Geometric conventionalism and Carnap's principle of tolerance," *Studies in History and Philosophy of Science* 25: 773–83.

(1995). "Tolerance and metalanguages in Carnap's *Logical Syntax of Language*," *Synthese* 103: 123–39.

(1999). "Tarski on 'essentially richer' metalanguages," *Journal of Philosophical Logic* 28: 1–28.

DeVidi, R. (1996). "Frege on Indexicals: Sense and Context Sensitivity," doctoral dissertation, University of Western Ontario.

DiSalle, R. (2002). "Conventionalism and modern physics: a re-assessment," *Noûs* 36: 169–200.

(2006). *Understanding Space-time: The Philosophical Development of Physics from Newton to Einstein* (Cambridge University Press).

Dummett, M. (1981). *Frege: Philosophy of Language* (Cambridge, MA: Harvard University Press, second edn).

(1982). "Frege and Kant on geometry," *Inquiry* 25: 233–54, reprinted in Dummett (1991c): 126–57.

(1991a). *Frege: Philosophy of Mathematics* (Cambridge, MA: Harvard University Press).

(1991b). *The Logical Basis of Metaphysics: The William James Lectures, 1976* (Cambridge, MA: Harvard University Press).

(1991c). *Frege and Other Philosophers* (Oxford University Press).

(1998). "Neo-Fregeans: in bad company?," in Schirn (1998: 369–88).

(2007). "Reply to Jan Dejnozka," in R. E. Auxier and L. E. Hahn (eds.), *The Philosophy of Michael Dummett* (Chicago and La Salle, IL: Open Court): 114–26.

Einstein, A. (1905). "On the electrodynamics of moving bodies," trans. W. Perrett and G. B. Jeffrey, reprinted in A. Einstein *et al.* (eds.), *The Principle of Relativity* (New York: Dover, 1952): 37–65.

(1921). "Address to the Prussian Academy," trans. S. Bargmann as "Geometry and experience," reprinted in A. Einstein, *Sidelights on Relativity* (New York: E. P. Dutton and Co., 1923): 27–45.

Enderton, H. (1972). *A Mathematical Introduction to Logic* (New York: Academic Press).

English, J. (1973). "Underdetermination: Craig and Ramsey," *Journal of Philosophy* 70: 453–61.

Feferman, S. (1986). "Introductory note to 1931c," in Feferman *et al.* (1986: 208–13).

Feferman, S. *et al.* (eds.) (1986). *Kurt Gödel: Collected Works*, vol. 1 (Oxford University Press).

(eds.) (1990). *Kurt Gödel: Collected Works*, vol. 11 (Oxford University Press).

Feigl, H. (1950). "Existential hypotheses," *Philosophy of Science* 17: 192–223.

Feynman, R. (1963). *Lectures on Physics*, vol. 1 (New York: Addison-Wesley).

Frapolli, M. J. (ed.) (2005). *F. P. Ramsey: Critical Reassessments* (London: Continuum Studies in British Philosophy).

Frege, G. (1874). Dissertation for the *venia docendi* in the Philosophical Faculty of Jena, trans. H. Kaal, as "Methods of calculation based on an extension of the concept of quantity," reprinted in McGuinness (1984: 56–92).

(1879). *Begriffsschrift: eine der arithmetischen nachgebildete Formelsprache des reinen Denkens*, trans. S. Bauer-Mengelberg as *Begriffsschrift: A Formula Language, Modeled upon that of Arithmetic, for Pure Thought*, reprinted in van Heijenoort (1967: 5–82).

(1880/1). "Booles rechnende Logik und die Begriffsschrift," in H. Hermes *et al.* (eds.), *Gottlob Frege: Nachgelassene Schriften* (Hamburg: Felix Meiner Verlag, 1983): 9–52, trans. P. White and R. Hargreaves as "Boole's logical calculus and the concept-script," in H. Hermes *et al.* (eds.), *Gottlob Frege: Posthumous Writings* (University of Chicago Press, 1980): 9–46.

(1884). *Die Grundlagen der Arithmetik*, trans. J. L. Austin as *The Foundations of Arithmetic: A Logico-Mathematical Enquiry into the Concept of Number* (Evanston: Northwestern University Press, 1980).

(1891). *Function and Concept*, trans. P. Geach, in McGuinness (1984: 137–56).

(1893–1903). *Grundgesetze der Arithmetik*, partially trans. M. Furth as *The Basic Laws of Arithmetic: Exposition of the System* (University of California Press, 1964).

(1897). "Logic," in H. Hermes *et al.* (eds.), *Gottlob Frege: Posthumous Writings* (University of Chicago Press, 1980): 126–51.

(1904). "What is a function?," trans. P. Geach, in McGuinness (1984: 285–91).

Freudenthal, H. (1962). "The main trends in the foundations of geometry in the 19th Century," in E. Nagel *et al.* (eds.), *Logic, Methodology and Philosophy of Science: Proceedings of the 1960 International Congress* (Stanford University Press): 613–21.

Friedman, M. (2001). *Dynamics of Reason* (Stanford: CSLI).

(2006). "Carnap and Quine: twentieth century echoes of Kant and Hume," *Philosophical Topics* 34: 35–58.

(2012). "Carnap's philosophical neutrality between realism and instrumentalism," in M. Frappier, D. Brown, and R. DiSalle (eds.), *Analysis and Interpretation in the Exact Sciences: Essays in Honour of William Demopoulos* (Dordrecht: Springer): 95–114.

Friedman, M. and R. Creath (eds.) (2007). *The Cambridge Companion to Carnap* (Cambridge University Press).

Gardner M. (1979). "Realism and instrumentalism in 19th century atomism," *Philosophy of Science* 46: 1–34.

Geach, P. (1976). "Critical notice of Michael Dummett, *Frege: Philosophy of Language*," *Mind* 85: 436–49.

Gödel, K. (1944). "Russell's mathematical logic," reprinted in Feferman *et al.* (1990: 119–41).

Goldfarb, W. (1979). "Logic in the twenties: the nature of the quantifier," *Journal of Symbolic Logic* 44: 351–68.

(1989). "Russell's reasons for ramification," in Savage and Anderson (1989: 24–40).

(1997). "Semantics in Carnap: a rejoinder to Alberto Coffa," *Philosophical Topics* 25: 51–66.

Goldfarb, W. and T. Ricketts (1992). "Carnap and the philosophy of mathematics," in D. Bell and W. Vossenkuhl (eds.), *Science and Subjectivity* (Berlin: Akademie Verlag): 61–78.

Grabiner, Judith (1981). *The Origins of Cauchy's Rigorous Calculus* (Cambridge, MA: MIT Press).

Gupta, A. (2006). *Empiricism and Experience* (Oxford University Press).

Gupta, A. and N. Belnap (1993). *The Revision Theory of Truth* (Cambridge, MA: MIT Press).

Hale, B. (1994). "Dummett's critique of Wright's attempt to resuscitate Frege," *Philosophia Mathematica* series III, 2: 122–47.

Hallett, M. (1984). *Cantorian Set Theory and Limitation of Size* (Oxford University Press).

(1990). "Physicalism, reduction and Hilbert," in A. D. Irvine (ed.), *Physicalism in Mathematics* (Dordrecht: Kluwer): 183–258.

Heck, R. (1995). "Definition by induction in Frege's *Grundgesetze der Arithmetik*," in Demopoulos (1995: 295–333).

Heck, R. and J. Stanley (1993). "Reply to Hintikka and Sandu: Frege and second-order logic," *Journal of Philosophy* 90: 416–24.

Hempel, C. G. (1958/1965). "The theoretician's dilemma: a study in the logic of theory construction," in H. Feigl and G. Maxwell (eds.), *Minnesota Studies in the Philosophy of Science*, vol. II (Minneapolis: University of Minnesota Press, 1958): 37–98, reprinted with minor alterations in C. G. Hempel, *Aspects of Scientific Explanation* (New York: The Free Press, 1965): 173–226. Page references are to the 1965 reprinting.

(1963). "Implications of Carnap's work for the philosophy of science," in Schilpp (1963: 685–710).

Hilbert, D. (1904). "On the foundations of mathematics," trans. S. Bauer-Mengelberg, reprinted in van Heijenoort (1967: 129–38).

(1925). "On the infinite," trans. S. Bauer-Mengelberg, reprinted in van Heijenoort (1967: 367–92).

Hintikka, J. and G. Sandu (1992). "The skeleton in Frege's cupboard: the standard versus nonstandard distinction," *Journal of Philosophy* 89: 290–315.

Hume, D. (1739). *Treatise of Human Nature*, ed. L. A. Selby-Bigge (Oxford University Press, 1888).

Isaacson, D. (1992). "Some considerations on arithmetical truth and the ω-rule," in M. Detlefsen (ed.) *Proof, Logic and Formalization* (London and New York: Routledge): 94–138.

Jeshion, R. (2001). "Frege's notions of self-evidence," *Mind* 110: 937–76.

Kant, I. (1787). *Kritik der reinen Vernunft*, trans. N. K. Smith as *Immanuel Kant's Critique of Pure Reason* (London: Macmillan, 1956).

Keenan, E. (2001). "Logical objects," in C. A. Anderson and M. Zeleny (eds.), *Logic, Meaning and Computation: Essays in Memory of Alonzo Church* (Dordrecht: Kluwer): 149–80.

Ketland, J. (2004). "Empirical adequacy and ramsification," *British Journal for the Philosophy of Science* 55: 287–300.

Kitcher, P. (1979). "Frege's epistemology," *Philosophical Review* 88: 235–63.

Klement, K. C. (2001). "Russell's paradox in Appendix B of *The Principles of Mathematics*: was Frege's response adequate?" *History and Philosophy of Logic* 22: 13–28.

(2002). *Frege and the Logic of Sense and Reference* (London: Routledge).

Lewis, D. (1970). "How to define theoretical terms," *Journal of Philosophy* 67: 427–45.

Lillehammer, H. and H. Mellor (eds.) (2005). *Ramsey's Legacy* (Oxford University Press).

Linsky, B. (1999). *Russell's Metaphysical Logic* (Stanford: CSLI).

Linsky, L. (1987). "Russell's no-classes theory of classes," in J. J. Thompson (ed.), *Being and Saying: Essays in Honor of Richard Cartwright* (Cambridge, MA: MIT): 21–40.

Maddy, P. (2001). "Naturalism: friends and foes," *Philosophical Perspectives* 15: 37–67.

(2008). *Second Philosophy: A Naturalistic Method* (Oxford University Press).

Maxwell, G. (1968). "Scientific methodology and the causal theory of perception," in I. Lakatos and A. Musgrave (eds.), *Problems in the Philosophy of Science* (Amsterdam: North-Holland): 148–60.

(1970). "Structural realism and the meaning of theoretical terms," in M. Radner and S. Winokur (eds.), *Analyses of Theories and Methods of Physics and Psychology*, Minnesota Studies in the Philosophy of Science, vol. iv (Minneapolis: University of Minnesota Press): 181–92.

McGuinness, B. (ed.) (1980). *Gottlob Frege: Philosophical and Mathematical Correspondence*, trans. H. Kaal (University of Chicago Press).

(ed.) (1984). *Gottlob Frege: Collected Papers in Mathematics, Logic and Philosophy* (Oxford: Basil Blackwell).

Mellor, D. H. (ed.) (1990). *F. P. Ramsey: Philosophical Papers* (Cambridge University Press).

Mulder, H. L. and B. F. B. van de Velde-Schlick (eds.) (1979). *Moritz Schlick: Philosophical Papers*, vol. ii (Dordrecht: Kluwer).

Myhill, J. (1958). "Problems arising in the formalization of intensional logic," *Logique et Analyse* 1: 78–83.

(1974). "The undefinability of the set of natural numbers in the ramified *Principia*," in G. Nakhnikian (ed.), *Bertrand Russell's Philosophy* (London: Duckworth): 19–27.

Nagel, E. (1961). *Structure of Science* (New York: Harcourt, Brace and World, Inc.).

Newman, M. H. A. (1928). "Mr Russell's causal theory of perception," *Mind* 37: 137–48.

Newman, W. (2006). "Max Newman: mathematician, codebreaker and computer pioneer," in B. J. Copeland (ed.), *Colossus: The Secrets of Bletchley Park's Codebreaking Computers* (Oxford University Press): 176–88.

Nye, M. J. (1972). *Molecular Reality: A Perspective on the Scientific Work of Jean Perrin* (New York: American Elsevier Inc.).

Parsons, C. (1964). "Frege's theory of number," reprinted with a postscript in C. Parsons, *Mathematics in Philosophy: Selected Essays* (Ithaca, NY: Cornell University Press, 1984): 150–75, and in Demopoulos (1995: 182–210).

(1990). "Introductory note to 1944," in Feferman *et al.* (1990: 102–18).

(2002). "Realism and the debate on impredicativity: 1917–1944," in W. Sieg *et al.* (eds.), *Reflections on the Foundations of Mathematics: Essays in Honor of Solomon Feferman* (London: A. K. Peters/CRC Press): 378–95.

Perrin, J. (1916). *Atoms*, authorized trans. D. L. Hammock (New York: Van Nostrand).

Poincaré, H. (1900). "Hypotheses in physics," in *The Foundations of Science* (Lancaster, PA: The Science Press, 1913): 127–39, authorized trans. by G. B. Halsted of "Les relations entre la physique expérimentale et la physique mathématique," *Revue Générale des Sciences Pures et Appliquées* 11: 1163–75.

(1912a). "Les conceptions nouvelles de la matière," *Foi et Vie* 15: 185–91. A translation of this essay by M. Frappier and J. Bub, together with an introduction by W. Demopoulos, M. Frappier, and J. Bub, appears in W. Demopoulos *et al.* (2012).

(1912b). "The relations between matter and ether," in H. Poincaré, *Mathematics and Science: Last Essays*, trans. John W. Bolduc (New York: Dover, 1963): 89–101.

Potter, M. (2000). *Reason's Nearest Kin: Philosophies of Arithmetic from Kant to Carnap* (Oxford University Press).

(2005). "Ramsey's transcendental argument," in Lillehammer and Mellor (2005: 71–82).

Psillos, S. (1999). *Scientific Realism: How Science Tracks Truth* (London: Routledge).

(2000). "Rudolf Carnap's 'Theoretical concepts in science,'" *Studies in History and Philosophy of Science* 31: 151–72.

Putnam, H. (1965). "Craig's theorem," *Journal of Philosophy* 62: 251–60.

(1977). "Realism and reason," *Proceedings and Addresses of the American Philosophical Association* 50: 483–98.

(1978). *Meaning and the Moral Sciences* (Boston: Routledge and Kegan Paul).

(1981). *Reason, Truth and History* (Cambridge University Press).

(2012). "A theorem of Craig's about Ramsey sentences," in M. DeCaro and D. Macarthur (eds.), *Philosophy in an Age of Science* (Cambridge, MA: Harvard University Press): 277–9.

Putnam, H. and P. Benacerraf (eds.) (1983). *Philosophy of Mathematics: Selected Readings* (Cambridge University Press, second edn).

Quine, W. V. O. (1951a). "Two dogmas of empiricism," reprinted with abridgements and expansions in Quine (1980: 20–46).

(1951b). "On Carnap's views on ontology," *Philosophical Studies* 2: 65–72.

(1963). "Carnap and logical truth," in Schilpp (1963: 385–406).

(1964). "Implicit definition sustained," *Journal of Philosophy* 61: 71–4.

(1969). "Ontological relativity," in *Ontological Relativity and Other Essays* (New York: Columbia University Press): 26–68.

(1980). *From a Logical Point of View: Logico-Philosophical Essays* (Cambridge, MA: Harvard University Press, second edn).

Ramsey, F. P. (1925a). "Epilogue," in Braithwaite (1931: 287–92).

(1925b). *The Foundations of Mathematics*, reprinted in Braithwaite (1931: 1–61) and in Mellor (1990: 164–224).

(1929). "Theories," in Braithwaite (1931: 212–36) and in Mellor (1990: 112–36).

Reck, E. (2004). "From Frege and Russell to Carnap: logic and logicism in the 1920s," in S. Awodey and G. Klein (eds.), *Carnap Brought Home* (La Salle, IL: Open Court): 151–80.

(2007). "Carnap and modern logic," in Friedman and Creath (2007: 176–99).

Ricketts, T. (1997). "Truth values and courses of values in Frege's *Grundgesetze*," in W. W. Tait (ed.), *Early Analytic Philosophy: Frege, Russell and Wittgenstein* (Chicago: Open Court): 187–211.

(2007). "Tolerance and logicism: logical syntax and the philosophy of mathematics," in Friedman and Creath (2007: 200–25).

Robinson, A. (1966). *Non-Standard Analysis* (Amsterdam: North-Holland).

Russ, S. B. (1980). "A translation of Bolzano's paper on the intermediate value theorem," *Historia Mathematica* 7: 152–85.

Russell, B. (1903). *The Principles of Mathematics* (New York: W.W. Norton, 1938 edn).

(1905). "On denoting," reprinted in Russell (1956: 41–56) and Russell (1994: 414–27).

(1908). "Mathematical logic as based on the theory of types," reprinted in Russell (1956: 59–102).

(1912). *The Problems of Philosophy* (Indianapolis, IN: Hackett, a reprint of the 1912 Home University Library edn).

(1914a). *Our Knowledge of the External World as a Field for Scientific Method in Philosophy* (London: Allen and Unwin).

(1914b). "The relation of sense-data to physics," reprinted in Russell (1929: 145–79).

(1919). *Introduction to Mathematical Philosophy* (London: Allen and Unwin).

(1924). *Logical Atomism*, reprinted in Russell (1956: 321–44).

(1927). *The Analysis of Matter* (New York: Dover, 1954 reprint).

(1929). *Mysticism and Logic* (New York: Norton).

(1948). *Human Knowledge: Its Scope and Limits* (London: Allen and Unwin).

(1956). *Logic and Knowledge: Essays 1901–1950*, ed. R. C. Marsh (London: Allen and Unwin).

(1959). *My Philosophical Development* (New York: Simon and Schuster).

(1968). *The Autobiography of Bertrand Russell*, vol. II (London: Allen and Unwin).

(1994). *The Collected Papers of Bertrand Russell*, vol. IV: *Foundations of Logic 1903–05*, the McMaster University edn, ed. A. Urquhart and A. C. Lewis (London and New York: Routledge).

Sandu, G. (2005). "Ramsey on the notion of arbitrary function," in Frapolli (2005: 237–56).

Savage, C. W. and C. A. Anderson (eds.) (1989). *Rereading Russell: Critical Essays on Bertrand Russell's Metaphysics and Epistemology*, Minnesota Studies in the Philosophy of Science, vol. XII (Minneapolis: University of Minnesota Press).

Schilpp, P. A. (ed.) (1963). *The Philosophy of Rudolf Carnap* (La Salle, IL: Open Court).

Schirn, M. (ed.) (1998). *Philosophy of Mathematics Today* (Oxford University Press).

Schlick, M. (1918/1925). *General Theory of Knowledge*, trans. A. E. Blumberg (La Salle, IL: Open Court, 1985).

(1926). "Experience, cognition and metaphysics," trans. P. Heath, reprinted in Mulder and van de Velde-Schlick (1979: 99–111).

(1932). *Three Lectures on Form and Content*, reprinted in Mulder and van de Velde-Schlick (1979: 285–369).

Simons, P. (1988). "Functional operations in Frege's *Begriffsschrift*," *History and Philosophy of Logic* 9: 35–42.

Stein, H. (1967). "Newtonian space-time," *Texas Quarterly* 10: 174–200.

(1989). "Yes, but ... Some skeptical remarks on realism and anti-realism," *Dialectica* 43: 47–65.

Sullivan, P. and M. Potter (1997). "Hale on Caesar," *Philosophia Mathematica* series III, 5: 135–52.

Swijtink, Z. G. (1976). "Eliminability in a cardinal," *Studia Logica* 35: 71–89.

Uebel, T. (2007). *Empiricism at the Crossroads: The Vienna Circle's Protocol-Sentence Debate* (Chicago and La Salle, IL: Open Court).

van Benthem, J. (1978). "Ramsey eliminability," *Studia Logica* 37: 321–36.

van Fraassen B. C. (1980). *The Scientific Image* (Oxford University Press).

(1989). *Laws and Symmetry* (Oxford University Press).

(1997). "Putnam's paradox: metaphysical realism revamped and evaded," *Philosophical Perspectives: Mind Causation and World* 11: 17–42.

(2008). *Scientific Representation: Paradoxes of Perspective* (Oxford University Press).

(2009). "The perils of Perrin, in the hands of philosophers," *Philosophical Studies* 143: 5–24.

van Heijenoort, J. (ed.) (1967). *From Frege to Gödel: A Sourcebook in Mathematical Logic, 1879–1931* (Cambridge, MA: Harvard University Press).

Wang, H. (1957). "The axiomatization of arithmetic," *Journal of Symbolic Logic* 22: 145–57.

Wehmeier, K. F. (2008). "Wittgensteinian tableaux, identity, and co-denotation," *Erkenntnis* 69: 363–76.

(2009). "On Ramsey's 'silly delusion' regarding *Tractatus* 5.53," in G. Primiero and S. Rahman (eds.), *Knowledge and Judgement: Essays in Honor of G. Sundholm* (London: College Publications): 353–68.

Whitehead, A. N. and B. Russell (1910). *Principia Mathematica*, vol. I (Cambridge University Press).

(1912). *Principia Mathematica*, vol. II (Cambridge University Press).

Wilson, M. (1992). "Frege: the royal road from geometry," reprinted with a post-script in Demopoulos (1995: 108–59).

(1993). "Honorable intensions," in S. Wagner and R. Warner (eds.), *Naturalism* (South Bend, IN: Notre Dame University Press): 53–94.

Winnie, J. (1967). "The implicit definition of theoretical terms," *British Journal for the Philosophy of Science* 18: 223–9.

(1970). "Theoretical analyticity," in R. Cohen and M. Wartofsky (eds.), *Boston Studies in the Philosophy of Science*, vol. VIII (Dordrecht: Reidel): 289–305.

(1977). "The causal theory of space-time," in J. Earman *et al.* (eds.), *Foundations of Space-Time Theories*, Minnesota Studies in the Philosophy of Science, vol. VIII (Minneapolis: University of Minnesota Press): 134–205.

Wittgenstein, L. (1922). *Tractatus Logico-Philosophicus*, trans. D. Pears and B. McGuinness (New York: Routledge Classics, 2001 edn).

Wright, C. (1983). *Frege's Conception of Numbers as Objects* (Aberdeen University Press).

(1990). "Field and Fregean Platonism," in A. Irvine (ed.), *Physicalism in Mathematics* (Dordrecht: Kluwer): 73–94.

(1997). "On the philosophical significance of Frege's theorem," in R. Heck (ed.), *Language, Thought, and Logic: Essays in Honour of Michael Dummett* (Oxford University Press): 201–44.

(1998a). "On the harmless impredicativity of N$^=$ ('Hume's principle')," in Schirn (1998: 339–68).

(1998b). "Response to Dummett," in Schirn (1998: 389–406).

Youschkevitch, A. P. (1976). "The concept of function up to the middle of the 19th century," *Archive for the History of Exact Science* 16: 37–85.

Index

a priori
  Frege's definition, 13–17
  traditional definition, 10–11, 174
absolute velocity, 37–9, 118–19, 138–9
abstraction principles, 206–9
abstractionism
  and Dedekind, 235–7
  Frege's critique, 177–8
  and spatio-temporal intuition, 176–8
  and Thomae, 177–8
acceleration, 118
Ainsworth, P., 152 n. 12
analysis
  contextual, 210–11, 228
  Frege on, 169–72, 174–6, 181–3
  nineteenth-century mathematical, 9, 11, 169,
    170, 171 n. 5, 171, 175, 182, 185
  and reconstruction, 19–21, 109–15, 140–2,
    148–9
  Russell's notion of, 109–11, 110 n. 2,
    148 n. 10, 148
  versus justification, 21–6
analytic truth, 28–45, 48
  and arithmetical truth in *Logical Syntax of
    Language*, 49–52
analyticity. *See also* Carnap's thesis; Tractarian
  strategy
  Carnap, 28–36, 43–5, 49–52, 159
  Frege, 8, 14, 17–18, 22, 29
  Ramsey, 242–4, 246
  Wittgenstein, 30–3, 242–3
Anderson, C. A., 136 n. 28
apriority
  of arithmetic, 8, 13, 15–21
argument independence, 210, 220–4
arithmetic
  application of, 19, 43, 89, 197–8
  generality of, 10–11, 12–13, 22, 27, 41, 42, 186,
    203–4
  and geometry, 28, 34–6, 42–3
  nonstandard, 74, 75–7, 143, 148–55

primitive recursive, 49–51
  and rigor, 9–12, 169–76, 185–6
  Robinson (the system Q), 23
arithmetical knowledge
  Frege versus Dedekind, 11–12, 178–80, 196–9,
    230–7, 240
  Frege versus Hilbert, 21–6
  Frege versus Whitehead and Russell, 19–21,
    191–2 n. 16†, 204–9, 213–14, 216–20, 226,
    230–7, 245–6, 252–4
arithmetical truth, 13, 29–36, 49–51. *See also*
  analytic truth
atomic hypothesis. *See also*
  realism–instrumentalism controversy
  and Einstein, 55–7, 59–62
  and Perrin, 55–7, 59–62, 79–81
  and Poincaré, 55 n. 10, 79–81
  and realism, 52–62, 78–81
Axiom of Choice, 20, 133, 192, 243, 246, 248–9,
  249 n. 12
Axiom of Infinity
  and Dedekind infinity, 213–14, 229
  in *Principia Mathematica*, 19–21, 107, 133, 207
  Ramsey's interpretation, 242, 243–52
Axiom of Reducibility, 211, 212–14, 220 n. 7,
  226–9
  and contingency, 229
  and Ramsey, 155–6, 212–13

bad company objections, 204, 207–9
Basic Law V, 22, 184, 209, 213 n. 3
Beck, L. W., 156 n. 17
Bell, J. L., 24 n. 16, 182 n. 22, 187 n. 5, 190 n. 15, 238
  n. 1, 238, 240–1, 248
Benacerraf, P., 169–70, 174 n. 10, 175–6
Berkeley, G., 111–12
Bernays, P., 47 n. 2, 47
Bolzano, B., 9, 11–12, 169–71, 171 n. 5, 174–5, 185
Boole, G., 9, 171
Boolos, G., 3, 20 n. 12, 23 n. 14, 133 n. 27, 178 n. 16,
  179 n. 17, 181, 187 n. 4, 187, 191 n. 16, 192, 207

Bottazzini, U., 171 n. 3, 182 n. 23
Braithwaite, R. B., 156–8, 162–3, 243 n. 8
Brownian motion, 55–6, 56 nn. 11–12, 60, 79
Bub, J., 79 n. 13
Burge, T., 13 n. 6, 16–17
Burgess, J., 24 n. 15, 238 n. 3, 239 n. 6

Cantor, G., 9, 169–71, 230
Cantor's theorem, 240
cardinal numbers
    Frege–Russell definition, 112, 131–3
    and von Neumann ordinals, 188–91
cardinality operator, 18–21, 185, 186, 191–2, 193–4,
    200–1, 204, 205, 206
    type-raising versus type-lowering, 21–2, 191–2
Carnap, R. *See also* analyticity; arithmetical truth;
    Carnap's thesis; internal questions; external
    questions; logicism; Ramsey-sentence
    reconstruction
    on analyticity, 28–45, 49–52
    on arithmetical truth, 28–30, 40–3,
        49–51
    Aufbau, 94–6, 104–7
    and conventionalism, 32 n. 6, 34, 36, 47 n. 3,
        51–2, 58, 59, 62, 74–5, 83
    empirical content, 83–4
    "Empiricism, semantics and ontology",
        46–67, 78†
    and Frege, 28–30, 40–3
    interpretation of the Ramsey sentence of a
        theory, 64–7, 153–8
    and linguistic frameworks, 46–67, 77–8
    Logical Syntax of Language, 29 n. 1, 29, 30 n. 2,
        31 n. 4, 41 n. 17, 44 n. 20, 49–52, 52 n. 8, 61
    and logicism, 29, 30–3, 245
    neutrality about ontology, 68, 71, 84–5
    observation and theoretical vocabulary, 62–7,
        63 n. 20, 73–81, 121–6, 140, 158–65
    partial interpretation, 122–6
    principle of tolerance, 28, 61, 75
    and Ramsey, 62–7, 71–8, 108–9, 121–6, 158–62,
        245–6
    realism–nominalism controversy, 47,
        51–2, 66
    reconstruction of the language of science, 53,
        62–7, 68, 71–8, 83, 84–5, 87, 91, 108, 121–6,
        142, 158–61
    and Russell, 95, 104–7, 123, 158–62
    and the structuralist thesis, 82–5, 161–2
    on theoretical analyticity, 121–2, 159 n. 18, 159
    and van Fraassen, 68–9, 71–3, 126–8
    and Wittgenstein, 30–6, 43–5, 107
Carnap sentence, 121–2
Carnot's principle, 55–6
Carnap's thesis, 28–45

Cartwright, R., 115 n. 8
Carus, A., 52 n. 8, 67
Cauchy, A.-L., 9, 169–71, 185
chains, 133 n. 25, 179 n. 18, 179, 184
Church, A., 230–2
class
    logical conception of, 211–14
    paradox, 211–13, 230, 233
    versus set, 216–26
comprehension scheme, 13 n. 8, 179, 225
concepts, 18–21, 26–7, 40–1, 181–4, 186–204,
    238. *See also* extension of a concept
    Fregean versus propositional functions, 230–7,
        238–42, 252–4
confirmationist theory of meaning, 43, 55
constructive empiricism, 68–73, 75, 78, 84 n. 20,
    85–9, 88 n. 23, 88 n. 26, 123, 126–8,
    150 n. 11. *See also* empiricist structuralism;
    van Fraassen
context principle, 192–5, 199, 202, 206
contextual definition, 19, 120, 185, 186–8, 187 n. 3,
    192–6, 198, 199–200, 201–3. *See also*
    Dummett; Frege criterion of identity;
    Hume's principle; Wright
    and bad company objections, 204, 206–9
    Dummett on, 204–6
    and impredicativity, 205
    Wright on, 204–5, 206–9
conventionalism, 32 n. 6, 35–7, 45, 47 n. 3, 52, 58,
    59–62, 83, 90, 100, 202, 208
Couturat, L., 180 n. 20
Craig, W., 146–7
Craig interpolation theorem, 116–17
Craig transcription, 142–6, 160
criterion of identity. *See also* contextual definition;
    Hume's principle
    and the epistemic status of geometry and
        arithmetic, 3, 28, 33–45
    and Frege's contextual definition, 18–19,
        185–209
    and Hume, 3–5, 18–24, 186–204
    for length, 39
    for number, 3, 5, 19, 40–2
    for time of occurrence, 37–9, 137–9

Dedekind, R. *See also* Frege–Dedekind definition
    of number
    categoricity theorem, 197, 235, 240
    and Fregean thoughts, 233–7
    on knowledge of number, 235–7
    his proof of an infinite system, 235–6
    and Russell's propositional paradox, 235–7
Dedekind infinity, 16, 19–22, 25, 27, 133, 187 n. 5,
    192, 204, 213, 214, 235, 237. *See also* Frege's
    theorem; Russell's theorem

Dedekind–Peano axioms, 13, 16–17, 19, 23, 24 n.
    16, 28, 29, 42, 131–2, 133, 187, 197 n. 22,
    240. *See also* Peano arithmetic;
    Peano–Dedekind model
DeVidi, R., 193 n. 17
Dirichlet, L., 182 n. 23, 182
DiSalle, R., 39 n. 13, 43 n. 19, 136 n. 29, 142 n. 2
Dummett, M., 13 n. 8, 40 n. 16, 171 n. 4, 171, 174
    n. 10, 177 n. 15, 189 n. 11–13, 196 n. 21, 204 n.
    25, 238–41
    and bad company objections, 204, 207–9
    on Dedekind, 235
    Frege: Philosophy of Language, 238, 239, 240
    Frege: Philosophy of Mathematics, 239 n. 5, 239,
        241
    on the interpretation of second-order
        quantifiers, 239
    and realism, 58
    on the vicious circularity of Hume's principle,
        204–6

Eddington, A., 90, 92 n. 7, 133
Einstein, A., 33, 46, 55–7, 59–61, 62, 66. *See also*
    Einsteinian strategy; special relativity
    and Poincaré, 36
    and the principle of free mobility, 36, 39
    on pure and applied mathematics, 33–6, 42–3
    on simultaneity, 36–9, 137–9, 141
Einsteinian strategy, 34–6, 44
empirical adequacy, 70, 72–3, 85–9, 126–8, 151–3
    of the atomic hypothesis, 80, 81
    and Ramsey sentences, 122, 125, 126–8
empiricism, 69, 165–8, 169. *See also* Carnap;
    Gupta; Ramsey; Russell; constructive
    empiricism
    and foundationalism, 109, 167
    and phenomenalism, 51, 90, 91–2, 94, 100, 105,
        155–7, 159, 160
empiricist structuralism, 68, 85–9. *See also*
    constructive empiricism; van Fraassen
English, J., 116
epistemic dependency thesis, 218–20, 226, 229
equinumerosity, 19, 22, 40–3
extension of a concept, 14–15, 17, 21, 22, 27, 183–4,
    189–91, 195–6, 198–9, 204–9, 217
external questions, 46–67, 78, 85 n. 21. *See also*
    internal questions; Carnap; linguistic
    frameworks

factual content, 30, 32, 43–5
    Carnap on, 28, 44, 46, 64, 67, 121,
        124, 158–61
    and the Ramsey sentence, 63–7, 72–3, 82, 121,
        124, 159–63
Feferman, S., 50 n. 7

Feigl, H., 63 n. 21
Feynman, R., 46
form–content distinction, 94, 95 n. 9, 100
foundationalism, 24 n. 16, 109, 181
Frapolli, M., 241
Frappier, M., 79 n. 13
Frege, G. *See also* Carnap; Frege arithmetic;
    Frege–Dedekind definition of number;
    Frege's theorem; Hilbert; logicism; Russell
    on abstraction, 178, 195–6
    on the analytic–synthetic distinction, 8, 10,
        13, 173
    on the application of arithmetic, 19, 27, 28, 30,
        40–3, 197–9, 203
    Begriffsschrift, 9–11, 27, 170–5, 177–84, 185, 203
    and Carnap, 28–30, 245–6
    concept of a function, 181–4, 233–7, 238–42,
        252–4
    contextual definition, 2, 19, 185, 186–8, 192–6,
        198–200, 201–2, 204–9
    criterion of identity, 2–4, 5, 18–21, 26–7, 28,
        40–1, 193–5, 197, 235, 237
    and Dedekind, 9, 11–12, 169–71, 173–6, 179,
        184, 196–8, 213–14, 235–7, 240
    definition of the ancestral of a relation, 178
    Grundgesetze, 25, 180, 183–4, 193, 196–7, 205,
        235, 239–41
    Grundlagen, 2–4, 8, 10, 13–15, 16, 17–19, 26, 40,
        112, 169, 173, 177–8, 180, 185–92, 193–7, 199,
        216, 246
    Julius Caesar problem, 195–6, 198, 202
    and Kant, 8–12, 13–14, 18, 26–7, 29, 173–5,
        181–4, 186, 188, 194, 203, 204
    on logical objects, 15, 16–18, 21, 199
    on numbers as objects, 14–15, 180, 188–204
    on rigor, 9–12, 169–84, 185–6
    and Russell–Whitehead, 19–21, 191–2, 213–14,
        216–17, 226, 229–37, 252–4
    on self-evidence, 9–10, 17–18, 21–2,
        172–3
    on spatio-temporal intuition, 9–12, 13–14, 18,
        171–8, 181–4, 185–6, 204
    theory of concepts and objects, 14–15, 183,
        188, 235
    theory of functions and classes, 226, 229,
        235, 253–4
    theory of number, 1, 3, 6, 19–21, 25, 188–91,
        197–9, 213, 217, 226
    theory of thoughts, 230–7
Frege arithmetic (FA), 18–19
    and Hilbert's program, 21–6
Frege–Dedekind definition of number, 18–20,
    185–9, 213–14, 217, 226, 229
Frege–Russell definition of the cardinal numbers,
    112, 131–3, 195

Fregean extensions, 14–15, 17, 21, 22, 189–91, 195–9, 204, 205, 206, 208, 217, 252–4
Frege's problem, 209
Frege's theorem, 18, 23, 185, 187–8, 192, 195, 203, 204, 205, 207
Friedman, M., 33 n. 7, 35 n. 11, 44 n. 20, 45 n. 21, 47 n. 3, 68 n. 1–2, 68, 71–3, 85, 142 n. 3, 150 n. 11, 152 n. 12
function dependence, 222–4

Gardner, M., 56 n. 13, 56
Geach, P., 23 n. 14, 23
geometry, 10–11, 90, 97–8, 118, 141, 163, 188, 203, 206
  application of, 28, 33–9
  and applied arithmetic, 28, 42–3
  epistemological status of, 3–5, 33
  Euclidean, 28, 34, 36–7, 93, 97, 129, 176
  Minkowskian, 28, 36–7, 42, 139, 141
Gödel, K., 224
  and Hilbert's program, 23–5
  incompleteness theorems, 23, 25, 49–51
  on Russell, 215, 223–4, 227
  "Russell's mathematical logic", 223, 224
Goldfarb, W., 33 n. 7, 33, 103 n. 15, 180 n. 20, 213 n. 3, 225
Gouy, L., 55, 56 n. 11
Grabiner, J., 171 n. 3
Gupta, A., 142 n. 3, 165–8, 201 n. 24

Hale, B., 204 n. 25
Hallett, M., 176 n. 14, 177 n. 15
Heck, R., 197 n. 22, 197, 240
Heisenberg, W., 90
Hempel, C., 62–5, 63 n. 20, 63 n. 21, 145–6, 159–60
Henkin, L., 128, 208, 209
Hilbert, D., 23–6, 25 n. 17, 82–3, 83 n. 18, 94, 140
Hintikka, J., 238–40, 241, 252–4
Hume, D.
  on the epistemic status of geometry and arithmetic, 3–5
  and the standard equality for number, 3, 4, 187 n. 3
Hume's principle, 3–4, 18–19, 20–4, 186–92, 193–4, 196, 197, 199, 204, 209. *See also* contextual definition; criterion of identity
  abstraction principles, 195, 207–9
  consistency of, 23–5, 186–7, 199
  logical form of, 21, 191

impredicativity
  and Hume's principle, 204–9
  and Russell's analysis of the paradoxes, 215, 223–6, 229

inductive class, 133
internal questions, 46–67
intuition
  and Frege, 8–12, 13–14, 18, 26–7, 29, 112, 170–8, 181–4, 185–6, 193–5, 216–17
  and Hilbert, 24–5
  and Kant, 10–11, 112–14, 173–4
Isaacson, D., 50 n. 7

Julius Caesar problem, 195–6, 198, 202, 204 n. 25

Keenan, E., 126 n. 22
Ketland, J., 142, 151–3, 152 n. 12
kinetic theory of gases, 56, 81
Kitcher, P., 174 n. 10

Leibniz, G., 90
Lewis, D., 123 n. 19
Lewis, G. N., 90
Linsky, B., 230 n. 15
Linsky, L., 213 n. 3
Lobachevsky, N., 182
Locke, J.
  on primary and secondary qualities, 93
  and Russell, 90–6
logic
  and concept formation, 5–6, 230–7
  Frege's development of, 8–12, 169–71, 185–6
  and intuition, 8–12, 13–14, 18, 26–7, 29, 170–8, 181–4, 185–6, 193–5, 216–17
  and mathematical reasoning, 9–12, 25–7, 29, 171–6, 179–84, 185–6
  and the principles of *Begriffsschrift*, 9–12, 13 n. 8, 171–84
  and rigor, 9–12, 169–76, 185–6
logical empiricism, 46–67, 68–89, 108–39, 140–68, 169
logical notion of class, 219, 223, 241
logical truth
  and the axioms of *Principia*, 211–14, 226–9
  and generality, 12, 242–3
  and Hume's principle, 199–202, 203
  and necessity, 12–13, 245–6
  and tautologousness, 30–2, 35, 242–3, 246
  and topic-neutrality, 200
logicism
  and the analysis of arithmetical knowledge, 8–27, 29–30, 185–204, 212–14, 226–9, 252–4
  and analyticity, 8–27, 29–36, 169–76
  and apriority, 8, 13, 15–21, 22–3, 25, 214
  Frege on, 8–27, 29–30, 169–84, 185–209, 252–4
  Frege versus Carnap on, 29–30
  and the generality of arithmetic, 13, 16, 19, 22, 27, 41, 204, 245–6

logicism (cont.)
and Hilbert's program, 24–6
and the justification of arithmetic, 9, 11, 13, 15–20, 21–6
and Kant, 8–12, 15–18, 26–7, 29, 172–6, 181–4, 185–8, 203–4
and logic, 169–84, 203–4, 211–14, 226–9
and neo-Fregeanism, 1, 204–9
nineteenth-century origins of, 9, 11, 169–71, 174, 185
Ramsey's transformation of, 238–54
Russell on, 19–21, 210–37, 242–4
and Russell's structuralism, 91–4, 112, 131–9
and Wittgenstein, 30–2, 31 n. 4, 107, 242–3, 245–6

McGuinness, B., 15 n. 9, 184 n. 25, 189 n. 12, 230 n. 14
Mach, E., 94
Maddy, P., 53, 59–62, 80, 85 n. 21
mathematical analysis
and Frege's conceptual notation, 8–12, 169–84
and rigor, 8–12, 169–76, 185–6
mathematical induction, 9, 10, 16, 133, 174, 180–1, 183, 186, 187 n. 5, 197 n. 22, 203–4. *See also* Myhill; Poincaré–Parsons objection
and the Axiom of Reducibility, 213–14
and the Frege–Dedekind definition of number, 25–6, 27, 171
Maxwell, G., 91 n. 4, 91, 101 n. 13, 101
Maxwell's demon, 56 n. 12
Maxwell's theory, 37, 138–9, 141
Mellor, H., 115 n. 9, 144 n. 6, 154 n. 13, 243 n. 8
Mill, J. S., 16
Myhill, J., 234

Nagel, E., 81 n. 16
Neurath, O., 167
Newman, M. H. A., 107
Newman's objection, 104–7, 161–8
and constructive empiricism, 70–1, 126–8
and Putnam's model-theoretic argument, 68 n. 2, 68, 70–1, 101–4, 122–6, 150 n. 11, 152 n. 12
to Russell, 96–101, 149–53
summarized, 150
Newtonian mechanics
and absolute simultaneity, 36–9, 42–3, 137–9
Ramsey on, 117–19
and special relativity, 37–9, 137–9, 141
and superfluous elements, 117–19
Nicod, J., 157
nominalism–realism controversy, 47, 52, 58 n. 14, 66

number. *See also* cardinal numbers
and cardinality, 19–21, 26–7, 40, 110 n. 2, 112, 131–2, 186–96, 199–204, 243–4
criterion of identity for, 19–21, 26–7, 28, 40–3, 193–4, 195, 197, 221, 235, 237
definition of, 19, 110 n. 2, 112, 189, 195, 196, 198–9, 213–14, 217, 226, 229
in *Principia Mathematica*, 19–21, 112, 133, 191–2, 195, 210–29, 238–54
recognition judgments involving, 2, 5, 19, 22, 26, 40–3, 193–4, 201
Nye, M. J., 56 n. 11

omega-rule (ω), 50–1

Parsons, C., 180 n. 21, 180, 217 n. 5, 227 n. 12, 227
Peano, G., 221 n. 8
Peano arithmetic. *See also* Dedekind–Peano axioms; Peano–Dedekind model
first-order (PA), 16–17, 23–4, 28, 29, 42, 131–3, 189, 197 n. 22
second-order (PA^2), 13 n. 8, 13, 18–21, 23–6, 187
Peano–Dedekind model, 24 n. 16, 131–3, 187
Perrin, J., 46, 55–7, 59–62, 66, 81
phenomenalism, 51, 90–2, 94, 100, 105, 155–7, 159, 160
Platonism
as an external question, 47, 51–2, 65–7
naive Platonism, 192–5, 202–3
and semantics, 77
Poincaré, H., 36. *See also* atomic hypothesis; Poincaré–Parsons objection
and conventionalism, 36
and Perrin, 55 n. 10, 79 n. 12–14, 79–81, 80 n. 15
Poincaré–Parsons objection, 180–1
Potter, M., 204 n. 25, 230 n. 16, 242 n. 7, 242, 244, 250
power set axiom, 106
primary and secondary qualities, 93
primary and secondary systems, 115–21, 122–6, 144–6, 153–7
principle of free mobility, 36–9
principle of tolerance, 28, 61, 75
proposition, 12, 30–2, 115–21, 165–8, 242–4. *See also* Russell: propositional paradox
Russellian, 109–15, 148 n. 10
propositional functions
extensional, 238–54
and Fregean concepts, 233–7, 252–4
and the logical notion of class, 223, 241, 247
in *Principia Mathematica*, 210–37
ramified type hierarchy of, 213–16
simple type hierarchy of, 214–16, 241
protocol-sentence debate, 44 n. 20, 167
Psillos, S., 63 n. 21, 121 n. 15, 159 n. 19, 159

psychologism, 18
Putnam, H.
　on Craig transcriptions and Ramsey sentences,
　　142–6, 143 n. 5, 145 n. 8
　model-theoretic argument, 68 n. 2, 68, 70–1,
　　86, 102–4, 122–6, 128–31, 150 n. 11, 152 n. 12
　and the observational–theoretical distinction,
　　122–3, 142–6

quantification
　and impredicativity, 178–9, 205, 224–6
　and ramification, 214–16
　and Wittgenstein's nested variable convention,
　　244–7
Quine, W. V. O., 32, 48 n. 5, 92 n. 6, 97, 102, 103
　on Carnap on analyticity, 28, 32–3
　his notion of centrality, 43–5
　"Two dogmas", 32

ramification, 155–6, 210, 211, 212–16, 223, 226–7,
　229, 232–3, 253
ramified type hierarchy, 214–16, 227
Ramsey, F. P.
　on the Axiom of Choice, 243, 246, 248–9, 249
　　n. 12
　on the Axiom of Infinity, 242, 243–52
　and extensional propositional functions,
　　238–54
　Foundations of Mathematics, 241–9
　on Newtonian space-time, 117–19
　"Theories" 109–121, 153–158
Ramsey's principle, 155–7. *See also* primary and
　secondary systems; Ramsey's principle;
　Ramsey sentence; Ramsey-sentence
　reconstruction
Ramsey sentence, 63 n. 20, 72 n. 6, 91–2, 104,
　115–21, 116 n. 10, 121 n. 16, 140–64
　and Craig transcriptions, 142–6
Ramsey-sentence reconstruction, 46, 62–7, 63 n.
　21, 68, 72–81, 84 n. 20, 91, 107 n. ‡, 121–6
ramsification, 91–2, 104–5, 115–21, 142–7, 153–8
real numbers, 11–12, 29, 43 n. 19, 97, 169, 170,
　182–3, 185
realism
　and the atomic hypothesis, 46, 53–67, 78–81
　and medieval realism, 52
　as a metaphysical and external question, 46–67
　and structural realism, 93–4, 104, 151–2
　and unobservables, 53–4, 64–7, 68–72, 77–89, 94
realism–instrumentalism controversy
　and the atomic hypothesis, 46, 53–67, 78–81
　Carnap on, 46–67, 68–89
　as an external question, 46–67
　and the reconstruction of empirical theories,
　　62–7, 68–89

and structure, 82–9, 93–4, 104, 151–2, 161–8
and unobservable entities, 53–4, 64–7, 68–72,
　77–89, 94
recognition judgments, 19, 26, 40–3, 193–5, 201
reflexive class, 20, 133
Reichenbach, H., 88–9, 90
Ricketts, T., 19
Riemann, B., 182 n. 23, 182
rigor
　and gap-freeness, 9–12, 171–6, 185–6
　and nineteenth-century mathematical analysis,
　　9, 11–12, 169, 182–4, 185–6
Robb, A. A., 90
Robinson, A., 143 n. 4
Robinson arithmetic (Q), 23
Robinson consistency theorem, 117 n. 11
Russ, S. B., 175 n. 11
Russell, B. *See also* class; Frege; Ramsey;
　Whitehead
　Analysis of Matter, 90–107, 95 n. 11, 111, 113 n. 7,
　　131, 132 n. 25
　on appearance and reality, 91, 93, 112
　on Berkelian idealism, 111–12
　his causal theory of perception, 68 n. 2, 90
　on contextual definition, 110 n. 2, 111, 120–1, 195
　on Dedekind infinity of the numbers, 19–21,
　　133, 191–2, 213–14, 229
　and Frege, 19–21, 110 n. 2, 112, 191–2, 213–14,
　　216–17, 221 n. 8, 230–7, 245–6, 252–4
　fundamental principle of analysis, 109–11,
　　119–20, 148
　incomplete symbols, 110 n. 2, 218–19
　Introduction to Mathematical Philosophy, 92, 95
　　n. 10, 95 n. 11, 113 n. 7, 155 n. 14
　knowledge by acquaintance and by description,
　　91, 109–15, 119–21, 148–9, 151, 161, 166, 212
　knowledge of things versus knowledge of
　　truths, 212, 228
　logicism, 131–2, 210–37, 245–6, 252–4
　and method of extensive abstraction, 148–53
　and M. H. A. Newman, 96–104, 150, 161–4
　Our Knowledge of the External World, 155
　and phenomenalism, 90, 91–2, 100, 105, 155–7
　Principia Mathematica, 19–21, 95, 107, 112, 133,
　　135, 192 n. *, 192, 210–37, 241–54
　Principles of Mathematics, 230
　on the problem of interpretation, 131–9
　Problems of Philosophy, 109 n. 1, 109, 227
　propositional paradox, 230–7
　and Ramsey, 91–2, 115–21, 141–2, 153–8, 238–54
　Russell's paradox, 211–13, 230, 233
　on structure, 91–4, 96–101, 148–53, 161–2
　theory of denoting concepts, 221–2
　theory of descriptions, 109–10, 115, 120–1,
　　210, 228

Russell, B. (cont.)
  theory of functions and classes, 210–37
  theory of propositional understanding, 108,
    109–15, 119, 120, 121, 131, 148, 149
  type theory, 156, 191–2, 205, 210–17, 226–37
  vicious circle principle, 210, 212, 223–4
  and Wittgenstein, 104, 107, 242–4
Russell's theorem, 20, 192

Sandu, G., 238 n. 2, 238 n. 3, 238–41, 249 n. 12,
    252–4
Schlick, M., 94, 95 n. 9
Schrödinger, E., 90
Second Law of Thermodynamics, 56
self-evidence, 9–10, 13, 17–18, 21–2, 171–3
semantic dependency thesis, 217–19, 224
semantic view of theories, 69–70, 71, 86, 126–31
set theory, 95, 97, 98, 104, 107, 203, 216, 240
  and the logical theory of classes, 211–14
Simons, P., 182 n. 22
simple type hierarchy, 133, 212, 214–16, 241, 253
simultaneity, 36–9, 38 n. 13, 41–3, 43 n. 19,
    137–9, 141
special relativity
  and Minkowskian geometry, 28, 36–7, 42
  and Newtonian mechanics, 37–9, 137–9, 141
  and simultaneity, 36–9, 42–3, 137–9, 141
Stanley, J., 240
Stein, H., 78 n. 10, 118 n. 12
structural realism, 93–5, 104, 151–2
structuralism
  in philosophy of mathematics, 82–3, 199–203
  Russell's, 91–4, 96, 100, 101–2, 111–14, 134–5,
    148–53
  and the theory of theories, 82–9, 90–107,
    111–14, 134–9, 140–2, 148–53, 153–8, 161–8
structuralist thesis, 82–5, 140–1, 161–6
Sullivan, P., 204 n. 25
Swijtink, Z., 73 n. 8, 142
syntactic view of theories, 88 n. 23, 130
synthetic a priori knowledge, 13, 29–30, 33,
    35, 227

Tarski, A., 50, 51
theoretical analyticity, 121–2, 159 n. 18, 159. *See also*
    Carnap sentence
theory of ideal gases, 124

theory of types, 29, 156, 191, 211, 212–17,
    229, 230–3
Thomae, C. J., 177
Tractarian strategy, 31–2, 35, 43–4
typical ambiguity, 216, 222, 226

Uebel, T., 58 n. 14, 167 n. 21

van Benthem, J., 70 n. 5, 73 n. 8, 142, 143, 146 n. 9,
    146–7, 150
van Fraassen, B., 68–73, 69 n. 4, 80–1, 83 n. 19, 85
    n. 22, 85–9, 88 n. 23, 126–31, 130 n. 24, 150 n.
    11, 151 n. 11, 153. *See also* constructive
    empiricism; empiricist structuralism;
    semantic view of theories
  The Scientific Image, 69–71, 88 n. 26
  Scientific Representation, 85–9
vicious circle principle, 210, 212, 223–6
  classification of types of, 223
von Neumann, J., 189–91, 199

Wang, H., 179 n. 18
Wehmeier, K., 34 n. 10, 244 n. 9
Weierstrass, K., 9, 169, 171
well-ordering theorem, 132 n. 25
Weyl, H., 118, 217 n. 5
Whitehead, A. N., 8, 20, 90, 103, 110 n. 2, 112 n. 5,
    132 n. 25, 133, 157, 191–2, 194–5, 210–11,
    220 n. 7, 238 n. 4, 248–9, 250
Wilson, M., 182 n. 23, 188 n. 8
Winnie, J., 97, 125, 126
Wittgenstein, L. *See also* Ramsey; Tractarian
    strategy
  his "big idea", 242
  and logicism, 30–3, 107, 242–4
  nested variable convention, 244 n. 9, 244, 245,
    246, 250
  on propositions of logic, 30–1, 242–3
  rejection of identity, 242–4
  Tractatus, 30–2, 33, 104, 107, 242–4, 245, 246,
    247
Wright, C., 1, 187, 204–9
  syntactic priority thesis, 194–5

Youschkevitch, A. P., 171 n. 3

Zermelo, E., 199, 230, 240